Selected Titles in This Series

17 **Olga Kharlampovich, Editor,** Summer school in group theory in Banff, 1996, 1999

16 **Alain Vincent, Editor,** Numerical methods in fluid mechanics, 1998

15 **François Lalonde, Editor,** Geometry, topology, and dynamics, 1998

14 **John Harnad and Alex Kasman, Editors,** The bispectral problem, 1998

13 **Michel C. Delfour, Editor,** Boundaries, interfaces, and transitions, 1998

12 **Peter C. Greiner, Victor Ivrii, Luis A. Seco, and Catherine Sulem, Editors,** Partial differential equations and their applications, 1997

11 **Luc Vinet, Editor,** Advances in mathematical sciences: CRM's 25 years, 1997

10 **Donald E. Knuth,** Stable marriage and its relation to other combinatorial problems: An introduction to the mathematical analysis of algorithms, 1997

9 **D. Levi, L. Vinet, and P. Winternitz, Editors,** Symmetries and integrability of difference equations, 1996

8 **J. Feldman, R. Froese, and L. M. Rosen, Editors,** Mathematical quantum theory II: Schrödinger operators, 1995

7 **J. Feldman, R. Froese, and L. M. Rosen, Editors,** Mathematical quantum theory I: Field theory and many-body theory, 1994

6 **Guido Mislin, Editor,** The Hilton Symposium 1993: Topics in topology and group theory, 1994

5 **D. A. Dawson, Editor,** Measure-valued processes, stochastic partial differential equations, and interacting systems, 1994

4 **Hershy Kisilevsky and M. Ram Murty, Editors,** Elliptic curves and related topics, 1994

3 **Rémi Vaillancourt and Andrei L. Smirnov, Editors,** Asymptotic methods in mechanics, 1993

2 **Philip D. Loewen,** Optimal control via nonsmooth analysis, 1993

1 **M. Ram Murty, Editor,** Theta functions: From the classical to the modern, 1993

Summer School in Group Theory in Banff, 1996

Volume 17

CRM PROCEEDINGS & LECTURE NOTES

Centre de Recherches Mathématiques
Université de Montréal

Summer School in Group Theory in Banff, 1996

Olga Kharlampovich
Editor

The Centre de Recherches Mathématiques (CRM) of the Université de Montréal was created in 1968 to promote research in pure and applied mathematics and related disciplines. Among its activities are special theme years, summer schools, workshops, postdoctoral programs, and publishing. The CRM is supported by the Université de Montréal, the Province of Québec (FCAR), and the Natural Sciences and Engineering Research Council of Canada. It is affiliated with the Institut des Sciences Mathématiques (ISM) of Montréal, whose constituent members are Concordia University, McGill University, the Université de Montréal, the Université du Québec à Montréal, and the Ecole Polytechnique. The CRM may be reached on the Web at www.crm.umontreal.ca.

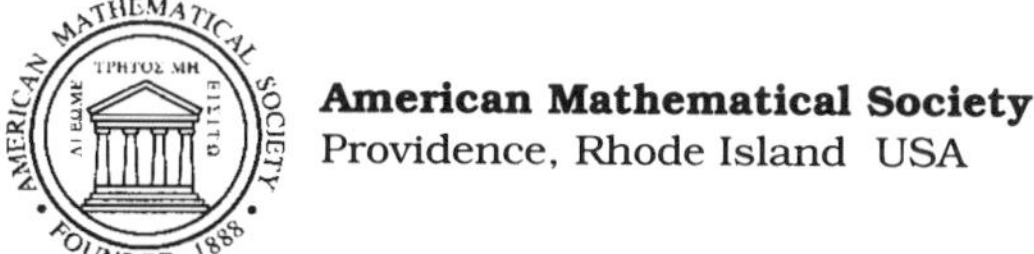

American Mathematical Society
Providence, Rhode Island USA

The production of this volume was supported in part by the Fonds pour la Formation de Chercheurs et l'Aide à la Recherche (Fonds FCAR) and the Natural Sciences and Engineering Research Council of Canada (NSERC).

1991 *Mathematics Subject Classification.* Primary 20E05, 20E06, 20F05, 20F06, 20F10, 20F32.

Library of Congress Cataloging-in-Publication Data

Summer School in Group Theory (3rd : 1996 : Banff, Alta.)
Summer School in Group Theory in Banff, 1996 / Olga Kharlampovich, editor.
p. cm. — (CRM proceedings & lecture notes, ISSN 1065-8580 ; v. 17)
"Third annual CRM Summer School took place in Banff (Alberta, Canada) from August 11 to 23 1996"—Pref.
Includes bibliographical references.
ISBN 0-8218-0948-2
1. Group theory—Congresses. I. Kharlampovich, Olga, 1958– . II. Title. III. Series.
QA174.S86 1996
512′.2—dc21 98-38712
CIP

∞ The paper used in this book is acid-free and falls within the guidelines
established to ensure permanence and durability.
This volume was typeset using AMS-TEX,
the American Mathematical Society's TEX macro system,
and submitted to the American Mathematical Society in camera ready
form by the Centre de Recherches Mathématiques.
Visit the AMS home page at URL: http://www.ams.org/

10 9 8 7 6 5 4 3 2 1   04 03 02 01 00 99

Contents

List of Participants ix

List of Speakers xi

Preface xiii

Some Open Problems
Gilbert Baumslag 1

Length Functions and Λ-trees
I. M. Chiswell 11

Introduction to Hyperbolic and Automatic Groups
S. M. Gersten 45

Description of Fully Residually Free Groups and Irreducible Affine Varieties Over a Free Group
O. Kharlampovich, A. Myasnikov 71

Homological Methods in Group Theory
Yuri Kuz′min 81

Pro-p Groups
Marcus du Sautoy 99

Identities of Representations of Finite Groups
Samuel M. Vovsi 131

List of Participants

Saeid Azam
University of Saskatchewan
Saskatoon, Saskatchewan (Canada)

Gilbert Baumslag
CUNY
New York, NY 10031-9100 (USA)

Nadia Benakli
Columbia University
New York, NY 10027-0029 (USA)

Oleg Bogoyavlenskij
Queen's University
Kingston, Ontario (CANADA)

Taoufik Bouzoubaa
Institut Henri Poincaré
Paris, Cedex 05 (FRANCE)

Frank Cannonito
University of California
Irvine, CA 92717 (USA)

Lisa J. Carbone
Columbia University
New York, NY 10027 (USA)

Ian Chiswell
Queen Mary & Westfield College
London E1 4NS (ENGLAND)

Ching-Fen Fuh
CUNY
New York, NY 10036 (USA)

Steve Gersten
University of Utah
Salt Lake City, UT 84112-1107 (USA)

Vincent Guirardel
Université Paul Sabatier
Toulouse, Cedex 31062 (FRANCE)

Kanta Gupta
University of Manitoba
Winnipeg, Manitoba (CANADA)

Narain Gupta
University of Manitoba
Winnipeg, Manitoba (CANADA)

Christian Hidber
ETH Zentrum
Zurich (SWITZERLAND)

Sergey Ivanov
University of Illinois
Urbana, Illinois 61801 (USA)

Laurie Johnson
State University of New York at Albany
Albany, NY12222 (USA)

Ilya Kapovich
CUNY
New York, NY 10031 (USA)

Olga Kharlampovich
McGill University
Montréal, Québec (CANADA)

Dongshin Kim
Univ. of Illinois at Urbana-Champaign
Urbana, IL 61801 (USA)

Dessislava Kochloukova
University of Cambridge
Cambridge CB2 1SB (ENGLAND)

Mailik Koubi
Université Paul Sabatier
Toulouse, Cedex 31062 (FRANCE)

Yuri Kuz'min
Moscow Univ. of Railway Transport Engineers
Moscow 101475 (RUSSIA)

John Labute
McGill University
Montréal, Québec, (CANADA)

Paul Libbrecht
Université du Québec à Montréal
Montréal, Québec (CANADA)

LIST OF PARTICIPANTS

Wan Lin
University of Manitoba
Winnipeg, Manitoba (CANADA)

Serguei Lioutikov
McGill University
Montréal, Québec (CANADA)

Ekaterina Lioutikova
McGill University
Montréal, Québec (CANADA)

Sal Liriano
City College
New York, NY 10031 (USA)

Natasa Macura
University of Utah
Salt Lake City, Utah 84112 (USA)

John Mighton
University of Toronto
Toronto, Ontario (CANADA)

Igor Mineyev
University of Utah
Salt Lake City, UT 84112 (USA)

Shahriar Mokhtari-Sharghi
Columbia University
New York, New York 10027 (USA)

Robert F. Morse
Eastern Nazarne College
Quincy, MA 02170 (USA)

Alexei Myasnikov
CUNY
New York, NY 10031-9100 (USA)

Igor Nikolaev
Institute of Mathematics
Moldavia, C.I.S. (?)

John O'neill
State Univ. of New York at Albany
Albany, NY 12222 (USA)

Akbay Rhemtulla
Univeristy of Alberta
Edmonton, Alberta (CANADA)

Luis Ribes
Carleton University
Ottawa, Ontario (CANADA)

Mark Sapir
University of Nebraska
Lincoln, NE 68588-0323 (USA)

Olga Sapir
University of Nebraska
Lincoln, NE 68588-0323 (USA)

Marcus du Sautoy
University of Cambridge
Cambridge, CB2 1SB (ENGLAND)

Jeremy Shor
Columbia University
New York, NY 10027 (USA)

Srilatha Singh
University of Michigan
Ann Arbor, MI 48105 (USA)

Shaobin Tan
Univ. of Saskatchewan
Saskatoon, Saskatchewan (CANADA)

Samuel Vovsi
Rutgers University
Princeton, NJ 08540 (USA)

Efim Zelmanov
Yale University
New Haven, CT 06520 (USA)

List of Speakers

Gilbert Baumslag,
City University of New York

Nadia Benakli,
Columbia University

Ian Chiswell
University of London

Marcus du Sautoy
Oxford University

Steve Gersten
University of Utah

Narain Gupta
University of Manitoba

Kanta Gupta
University of Manitoba

Sergey Ivanov
University of Illinois at Urbana

Olga Kharlampovich
McGill University

Yuri Kuz'min
Moscow University of Rail Trans. Eng.

John Labute,
McGill University

Alexei Myasnikov
City University of New York

Mark Sapir
University of Nebraska

S. M. Vovsi,
Rutgers University

Efim Zelmanov
Yale University

Preface

The third annual CRM Summer School took place in Banff (Alberta, Canada) from August 11 to 23, 1996. The scientific committee consisted of G. Baumslag (CUNY), D. Gildenhuys (McGill), O. Kharlampovich (McGill), E. Zelmanov (Yale).

The Summer School was primarily aimed at PhD students in their final years and recent PhDs.

The purpose of the school was also to bring together specialists, approaching group theory from different angles.

One of these is the geometric approach. Much of combinatorial group theory has its roots in geometrical and topological ideas. For example, the notion of non-positive curvature, implicitly underlying small-cancellation theory, has been abstracted and exploited by Gromov, Cannon and Rips, among others, resulting in a sizable body of work on hyperbolic (or negatively curved) groups. This work has many links to other areas like hyperbolic metric spaces, hyperbolic manifolds and group actions on R-trees. In this context, we should mention as well the geometric methods based upon Van Kampen diagrams. These methods were substantially developed by Ol'shanskii in his work on the Burnside, and related, problems.

A second approach is via logical methods. The area of algorithmic solvability of problems in group theory fall under this heading. Razborov (1990 Nevalinna prize winner) has constructed an algorithm which recognizes the solvability, or otherwise, of a system of equations over a free group. His construction uses ideas due to Makanin. The same ideas turned out to be very fruitful in many other investigations, in particular in Rips' solution of the Morgan-Shalen conjecture about groups acting on R-trees.

A third approach to combinatorial group theory has its roots in the Bass-Serre theory of groups acting on trees.

A big part of the lecture notes are contained in this book of the CRM Proceedings and Lecture Note Series. The contribution of Efim Zelmanov is to be published in a separate volume in the CRM Monograph Series. We are very grateful to the invited speakers and participants who have contributed to the books. We are very grateful to Louise Letendre and Andre Montpetit for their very professional expert work as well as Louis Pelletier for his great organisational support.

Finally it is a pleasure to thank the CRM and its director, Luc Vinet, for the opportunity to organize this School, the funding of the activities and the invitation to publish the resulting notes and papers in the CRM Series.

Olga Kharlampovich
Montreal, June, 1998

Centre de Recherches Mathématiques
CRM Proceedings and Lecture Notes
Volume 17, 1999

Some Open Problems

Gilbert Baumslag

1. General Remarks

My objective here is to discuss a number of problems connected to one-relator groups. The recent developments of hyperbolic and automatic groups have begun to play an increasingly important part in combinatorial group theory and they will play a supporting, not a leading, role here. I have elected to organize the problems that I am going to put forward in this note around another recent development, namely the ongoing attempt to carry over into group theory various elementary notions of algebraic geometry, which have been talked about in this conference and many other places in the last few years, by Alexei Myasnikov. Some of the ideas involved go back to V. S. Guba [**19**], to J. R. Stallings [**37**] and, perhaps also to I. Rips and to V. N. Remeslennikov [**34**]. The general theory of this algebraic geometry over groups has been under construction, so to speak, for a few years now and is the subject of a number of papers by Myasnikov, Remeslennikov and myself, two of which are as yet unpublished [**9–11**].

2. Preliminary Notions

I want, at the outset, to record some of the notation, some definitions and some theorems that will be used here.

A group G is termed *Hopfian* if $G/N \cong G$ implies that $N = 1$; otherwise not Hopfian or *non-Hopfian*. It is *linear* if it has a faithful matrix representation over a field and *residually finite* if its subgroups of finite index intersect in the identity. In this connection, let me point out that A. I. Malcev [**28**] has proved that finitely generated linear groups are residually finite and that finitely generated residually finite groups are Hopfian.

I will adopt the customary notation for commutators and conjugation

$$[x, y] = x^{-1}y^{-1}xy, \quad x^y = y^{-1}xy,$$

where x and y are elements in some group. The n-th term of the lower central series of a group G will be denoted by $\gamma_n(G)$, with $\gamma_1(G) = G$.

A finitely presented group G is termed *hyperbolic* if it has a finite presentation

$$G = \langle_1, \dots, x_m; r_1, \dots, r_n\rangle,$$

1991 *Mathematics Subject Classification.* Primary: 20F05.
This is the final form of the paper.

where the set of defining relators is closed under inversion and cyclic permutations, such that every word w which takes on the value 1 in G contains more than half a relator. Thus such a group has the property that every relator of length n can be expressed as a product of linearly many conjugates of the defining relations (see also [**17**]). The minimum number of such conjugates in the general case of a finitely presented group gives rise to a so-called *isoperimetric function*, which measures the complexity of the solution of the word problem for the group involved. Finally, a group is termed *automatic* if it has a regular language such that every pair of paths in its Cayley graph which end no more than a distance one apart, never stray more than a bounded distance from each other. Hyperbolic groups are automatic. We refer the reader to the book by D. B. A. Epstein [**15**] for a general reference to automatic groups and to the paper by M. Gromov [**18**], where the theory of hyperbolic groups was founded.

3. Test Groups

The problems that I am going to propose here can all be tested against a number of specific groups to see how the land lies. I have chosen to record five families of groups at this time and will discuss them more when they arise.

Here I will assume that l, m, n are positive integers with $l > 1$.

The first family consists of the so-called *BS-groups*, introduced way back in 1962 by Donald Solitar and myself [**13**]:

$$B_{m,n} = \langle a, t; a^{mt} = a^n \rangle. \tag{3.1}$$

The next family consists of one-relator groups with torsion, which are defined in a similar way.

$$B_{m,n,l} = \langle a, t; (a^{mt}a^{-n})^l \rangle. \tag{3.2}$$

The origin of this class (3.2) goes back some 30 years to the paper [**3**], where they were introduced to bolster an assertion made at that time, that one-relator groups with torsion are prone to less pathology than torsion-free one-relator groups. This has been borne out to a great extent over the years and I will reinforce this assertion with a new result about them in due course.

The next family consists of one-relator groups which are as far from being residually finite as possible [**6**]:

$$C_{m,n} = \langle a, t; a^{ma^t} = a^n \rangle. \tag{3.3}$$

Actually only $C_{1,2}$ has appeared in the literature [**6**] as of now. It was another example conjured up to demonstrate the complexity of one-relator groups. The family $C_{m,n}$ is the obvious generalisation of this group. And, like it, if $m = n + 1$, all of the finite quotients of $C_{m,n}$ are cyclic.

The next family (3.4) of groups, is the natural extension of the family (3.3):

$$C_{m,n,l} = \langle a, t; (a^{ma^t}a^{-n})^l \rangle. \tag{3.4}$$

This family is new, designed, like the others, to test some of the problems that I will discuss shortly.

Finally, a rather different family, which share many of the properties of free groups and may well be worth studying in detail:

$$G_w = \langle s, t, a_1, \dots, a_q; a_1 = ws^{-1}t^{-1}st \rangle. \tag{3.5}$$

Each of the groups given by (3.5) depends on a parameter w, which is a word in the given generators that involves a_1, but not s. I will discuss them and their relevance to the matter at hand, at the end of this note (cf. [**7**]).

4. Algebraic Geometry Over Groups

As I have already indicated, much of what I am going to say here involves algebraic geometry over groups. I have chosen to single out only a special case of one concept here, that of an *equationally Noetherian* group, introduced in [**9**] (see also [**10**]), where details of this and other notions are defined in complete detail and generality. To this end, let G be a given group, let F_n be the free group, freely generated by $x_1, \ldots, x_n$ and let

$$G[x_1, \ldots, x_n] = G * F_n$$

be the free product of G and F_n. I will use functional notation here, denoting an element $f \in G_n$ by

$$f = f(x_1, \ldots, x_n) = f(x_1, \ldots, x_n, g_1, \ldots, g_m) \tag{4.1}$$

thereby expressing the fact that the word representing f in $G[x_1, \ldots, x_n]$ involves the *variables* $x_1, \ldots, x_n$ and, as needed, the *constants* $g_1, \ldots, g_m \in G$. An n-tuple

$$v = (v_1, \ldots, v_n) \quad (v_i \in G) \tag{4.2}$$

is said to be a *root* of f if

$$f(v) = f(v_1, \ldots, v_n, g_1, \ldots, g_m) = 1.$$

If S is a subset of $G[x_1, \ldots, x_n]$ then v is said to be a root of S if it is a root of every $f \in S$. Let S be a subset of $G[x_1, \ldots, x_n]$. The *algebraic set over G defined by S*, or, more simply, the *algebraic set defined by S*, is, by definition, the set $V(S)$ of all roots of S:

$$V(S) = \{v = (v_1, \ldots, v_n) \mid v_i \in G, f(v) = 1, \text{ for all } f \in S\}.$$

The group G is said to be *equationally Noetherian* if for every choice of the integer n and every subset S of $G[x_1, \ldots, x_n]$

$$V(S) = V(S_0),$$

where S_0 is a finite subset of S. The class of equationally Noetherian groups is closed under subgroups and finite direct products. Furthermore, if a group G is equationally Noetherian and N is a finite normal subgroup of G, then G/N is also equationally Noetherian and if G is of finite index in the group H, then H is also equationally Noetherian. The latter statement constitutes two theorems of Vitali Romankov, Alexei Myasnikov and myself [**12**]. More important here is the observation that every linear group is equationally Noetherian. This was proved by by V. Guba [**19**], in the case of a free group and his proof immediately extends to linear groups [**10**]. It follows that groups which are not Noetherian are not linear.

Consider now the the family $B_{m,n}$ of BS-groups. Then [**10**]

Theorem 1. *$B_{m,n}$ is equationally Noetherian if and only if either $m = 1$ or $n = 1$ or $m = n$.*

It follows, in particular, that $B_{2,3}$ is not linear. There is another reason why those $B(m,n)$ which are not Noetherian are not linear. As I noted earlier, finitely generated linear groups are residually finite [**28**]. Now the $B_{m,n}$ which are not equationally Noetherian contain finitely generated subgroups which are non-Hopfian and hence not residually finite [**28**]. I will touch on this remark again later.

It is instructive to see why, for example,

$$H(= B_{2,3}) = \langle a, t; t^{-1}a^2t = a^3\rangle$$

is not equationally Noetherian.

It follows from Magnus' solution of the word problem for one-relator groups [**25**], that

$$[a, a^{2^n t^{-1}}] \neq 1.$$

Now observe that

$$(a^{3^n})^{t^{-n}} = a^{2^n}.$$

It follows that if

$$S = \{[x^{y^{-n}}, z] \mid n = 1, 2, \dots\}$$

then

$$x = a^{3^n}, \quad y = t, \quad z = a$$

is a root of

$$\{[x^{y^{-i}}, z] \mid i = 1, 2, \dots, n\}$$

but not a root of S. Hence H is not equationally Noetherian.

The presence of BS-groups as subgroups of one-relator groups groups are often an indication of pathology, so to speak, prompting me to put forward the first of my problems.

PROBLEM 1. Is every one-relator group which does not contain a BS-group equationally Noetherian? In particular, is every one-relator group with torsion equationally Noetherian?

One-relator groups with torsion are hyperbolic [**8**]. This follows from the fact that one can solve the word problem for them in linearly many steps, because of a theorem of B. B. Newman [**33**].

This suggests then the next problem.

PROBLEM 2. Is every hyperbolic group equationally Noetherian?

The notion of an exponential group was introduced by Roger Lyndon around 1960 and has been the subject of a good deal of investigation by Alexei Myasnikov and Vladimir Remeslennikov [**29, 30**]. In particular, one can form a number of so-called completions of groups so that their elements admit exponentiation by elements from a ring, for example $Z[x]$ and Q. If one completes a free group one obtains in the first instance the free $Z[x]$-groups studied first by R. C. Lyndon [**22**] and in the second the free Q-groups, which I investigated in my thesis [**1**]. Recently, using residual properties, the following theorem was proved in [**11**]:

THEOREM 2. *Free $Z[x]$-groups are equationally Noetherian.*

Here I would like to ask whether the corresponding result holds for free Q-groups.

PROBLEM 3. Are free Q-groups equationally Noetherian?

Actually the proper formulation of this problem involves the analogue in the category of Q-groups, i.e., uniquely divisible groups, of this notion of equationally Noetherian formulated here in the category of groups. But I will not go into this here.

I want next to turn my attention to some very, very old problems about one-relator groups.

5. Some Very, Very Old Problems

The word problem for one-relator groups was proved to be solvable in 1930 by W. Magnus [**25**], who also proved at about the same time, his celebrated freiheitssatz [**24**]. I have already alluded to this and will not concern myself with it further. I have also alluded to one measure of the complexity of the solution of the word problem, involving the number of uses of the defining relators needed to deduce that a given relator of a given length is the the identity. The group $G_{1,2}$ turns out to exhibit extraordinary complexity in this regard—this was proved by S. M. Gersten [**16**] 10 years ago. Precisely how complicated this notion of complexity for a one-relator group can be is not yet completely understood.

The other classical problems, the conjugacy and isomorphism problems, are still open. Arye Juhash has made some inroads into the conjugacy problem, which was solved for one-relator groups with torsion by Bill Newman some 20 years ago. However the general problem remains open. The isomorphism problem seems extremely difficult and little is known in general.

Zlil Sela [**35, 36**] has recently solved the isomorphism problem for torsion-free hyperbolic groups. In addition, Sela has proved that torsion-free hyperbolic groups are Hopfian. Although his methods do not as yet carry over to hyperbolic groups with torsion it is possible to deduce from the torsion-free case, that one-relator groups with torsion are Hopfian.

THEOREM 3. *Let G be a finitely generated, one-relator group with a non-trivial element of finite order. Then G is Hopfian.*

PROOF. There exists a normal, torsion-free subgroup N of G of finite index [**21**]. Let V be the variety defined by the finite group G/N (see [**32**]). Now the finitely generated groups in a variety generated by a finite group are finite [**31**] (see [**32**, p. 18]). So the verbal subgroup $V(G)$ of G defined by V is contained in N and is of finite index in G. Since $V(G)$ is fully invariant, every surjective endomorphism ϕ of G maps $V(G)$ into itself. It follows from the fact that $V(G)$ is of finite index, that $V(G)\phi = V(G)$. Now a subgroup of finite index in an hyperbolic group is again hyperbolic [**18**]. So $V(G)$ is hyperbolic and consequently, by Sela's theorem, it is Hopfian. Thus ϕ, restricted to $V(G)$, is monic. But $G/V(G)$ is finite and so ϕ is monic on $G/V(G)$. This means that ϕ is monic, i.e., an automorphism. This completes the proof. □

This theorem substantiates the remark that I made earlier about the nature of one-relator groups with torsion. The following problem underlines what it is that I have in mind for such groups.

PROBLEM 4. Are one-relator groups with non-trivial torsion

1. residually finite?
2. linear?

3. virtually free-by-cyclic?

The various families $B_{m,n,l}$ and $C_{m,n,l}$ provide a good collection of test cases for all parts of this problem. I believe that all of the groups $B_{m,n,l}$ have now been shown to be residually finite. I know that $C_{1,2,2}$ is also residually finite, but I do not know what the answers are to the other remaining questions for these groups. All of these families are interesting and I know very little about them. I will say a little more about them later.

Now suppose that

$$G = \langle X; s^n \rangle$$

is a one-relator group defined by the relator s^n, where s is not a proper power in the ambient free group in which it lies. Then it turns out that s is of order n in G [**27**], that the normal closure in G of sK is a free product of conjugates of the cyclic group generated by s [**14**] and that G/K is a torsion-free one-relator group defined by the single relator s. It is not at all clear what the relevance is to G of the properties of G/K. In order to focus more on the nature of one-relator groups, I would like to put forward the following problem, which highlights some aspects of the previous problem.

PROBLEM 5. Let

$$G = \langle X; s^n \rangle$$

and K be as above and put $H = G/K$.

1. Is G residually finite whenever H is?
2. Is G linear whenever H is linear?
3. Is G virtually free-by-cyclic whenever H is virtually free-by- cyclic?

6. Variations on Some Themes of Long Ago

Let me consider again the groups

$$B_{m,n} = \langle a, t; a^{mt} = a^n \rangle \quad (m, n > 0).$$

They divide naturally into two classes—the ones which are equationally Noetherian ones and the ones which are not. I have already described which ones are which. The equationally Noetherian ones in this class are simply the linear ones. I want to take the time now to look in some detail at the structure of one of them, specifically $B_{1,2} = H$, say. As I have already noted, H is linear. Indeed if we set

$$t = \begin{pmatrix} 1 & 0 \\ 0 & 2 \end{pmatrix}, \quad a = \begin{pmatrix} 1 & 0 \\ 1 & 1 \end{pmatrix}$$

then this defines a faithful representation of H in $\mathrm{GL}(2, Q)$. This makes it easy to understand the normal closure N of a in H. If we simply compute the conjugates of a under the powers of t, we find that N is a union of cyclic groups, indeed isomorphic to the additive group of Q of rational numbers whose denominators are powers of two. It is actually better to follow Magnus' original procedure and find a presentation of N. To this end, let me put

$$a_i = t^{-i} a t^i \quad (i \in \mathbb{Z}).$$

Then N can be presented as follows:

$$N = \langle \ldots, a_i, a_{i+1}, \ldots; \ldots, a_i^2 = a_{i+1}, a_{i+1}^2 = a_{i+2}, \ldots \rangle.$$

It follows that consecutive elements a_i, a_{i+1} in the sequence generate the one-relator group

$$\langle a_i, a_{i+1}; (a_i)^2 = a_{i+1} \rangle,$$

which is clearly infinite cyclic. It follows thence that N is a union of infinite cyclic groups. Notice that that there is an obvious automorphism of N, the one that shifts the a_i along, moving a_i to a_{i+1}. H is simply the semi-direct product of N and the infinite cyclic group on t, with t acting as this shift automorphism.

This description admits considerable variation. For example, let N now be the group presented as follows:

$$N = \langle \ldots, a_i, a_{i+1}, \ldots; \ldots, a_i^3 = a_{i+1}^2, a_{i+1}^3 = a_{i+2}^2, \ldots \rangle.$$

In this case N is not Abelian, since the a_is do not commute with each other. Consider the effect of adding the relations

$$\ldots, \quad [a_i, a_{i+1}] = 1, \quad [a_{i+1}, a_{i+2}] = 1, \quad \ldots$$

to N:

$$N' = \langle \ldots, a_i, a_{i+1}, \ldots; \ldots, a_i^3 = a_{i+1}^2, a_{i+1}^3 = a_{i+2}^2, \ldots, [a_i, a_{i+1}] = 1, [a_{i+1}, a_{i+2}] = 1, \ldots \rangle.$$

It is easy to see that if we put

$$\text{(6.1)} \qquad \ldots, \quad c_i = a_i^{-1} a_{i+1}, \quad c_{i+1} = a_{i+1}^{-1} a_{i+2}, \quad \ldots,$$

then $c_i^2 = c_{i+1}^3$; indeed the c_i satisfy precisely the same relations as the a_i, i.e.,

$$N' = \langle \ldots, c_i, c_{i+1}, \ldots; \ldots, c_i^3 = c_{i+1}^2, c_{i+1}^3 = c_{i+2}^2, \ldots \rangle.$$

So N' is isomorphic to N. This means that N is non-Hopfian. If we now extend N by the infinite cyclic group t acting on N by translation, we find that there is a surjection of the resultant group H to the group built up from N' in the same way. Since this surjection has a non-trivial kernel, it follows that H is non-Hopfian as well. It is not hard to see that this group H is simply $B_{2,3}$ in "expanded" form:

$$H = \langle a, t; a^{2t} = a^3 \rangle.$$

This discussion of H carries over also to the groups $C_{m,n}$:

THEOREM 4. *The groups $C_{m,n}$, where m and n are distinct primes, have the following properties*:

1. *$G_{m,n}$ is not Hopfian*;
2. *All of the finite factor groups of $G_{m,n}$ are cyclic.*

It is not at all clear how extensive the class of non-Hopfian one-relator groups is. Indeed I do not know of any non-Hopfian one-relator groups which are essentially different from BS-groups. In the hope of adding to our understanding of one-relator groups I would like to propose some further problems.

PROBLEM 6. Let G be a one-relator group presented on n generators and one defining relator r. If r lies in the derived group of the ambient free group, i.e., if the abelianisation of G is free abelian of rank n, is G Hopfian?

PROBLEM 7. Let G be a one-relator group. If G does not contain any BS groups, is G Hopfian?

PROBLEM 8. Is the group $C_{1,2}$ Hopfian?

Problem 9. If
$$G = \langle x_1, \ldots, x_n; [u,v] = 1\rangle$$
is G Hopfian? Residually finite? Automatic?

7. Parafree Groups

In conclusion I want to draw your attention to the groups G_w that I introduced at the outset. With this in mind, let me recall the

Definition 1. A group G is termed *parafree* if G is residually nilpotent and there exists a free group F such that
$$G/\gamma_n(G) \cong F/\gamma_n(F) \quad \text{for all } n.$$

Parafree groups, which can be likened to free groups, exist in profusion (see, e.g., [**2, 4, 5**]).

Theorem 5. *Let F be the free group on s, t, $a_1, \ldots, a_q$ and let w be an element of F which involves a_1 and does not involve s. In addition suppose that w lies in the k-th term $F^{(k)}$ of the derived series of F. Then G_w is a parafree with the additional property that*
$$G_w/G_w^{(k)} \cong H/H^{(k)},$$
where H is a free group of rank $q+1$.

Since parafree groups are so close to free groups, this suggest some obvious problems.

Problem 10. Are the G_w equationally Noetherian? Are all parafree groups equationally Noetherian? Are the finitely generated parafree groups automatic? Hyperbolic?

References

1. G. Baumslag, *Some aspects of groups with unique roots*, Acta Math. **104** (1960), 217–303.
2. ———, *Groups with the same lower central sequence as a relatively free group*. I. *The groups*, Trans. Amer. Math. Soc. **129** (1967), 308–321.
3. ———, *Residually finite one-relator groups*, Bull. Amer. Math. Soc. **73** (1967), 618–620.
4. ———, *More groups that are just about free*, Bull. Amer. Math. Soc. **74** (1968), 752–754.
5. ———, *Groups with the same lower central sequence as a relatively free group*. II. *Properties*, Trans. Amer. Math. Soc. **142** (1969), 507–538.
6. ———, *A non-cyclic one-relator group all of whose finite quotients are cyclic*, J. Austral. Math. Soc. **10** (1969), 497–498.
7. ———, *Musings on Magnus*, The Mathematical Legacy of Wilhelm Magnus: Groups, Geometry and Special Functions (Brooklyn, NY, 1992) (W. Abikoff, J. S. Birman, and K. Kuiken, eds.), Contemp. Math., vol. 169, Amer. Math. Soc., Providence, RI, 1994, pp. 99–106.
8. G. Baumslag, S. M. Gersten, M. Shapiro, and H. Short, *Automatic groups and amalgams*, J. Pure Appl. Algebra **76** (1991), no. 3, 229–316.
9. G. Baumslag, A. G. Myasnikov, and V. N. Remeslennikov, *Residually hyperbolic groups*, Russian Academy of Sciences (Siberian Division) (1995), 1–37.
10. ———, *Algebraic geometry over groups* (in preparation).
11. ———, *Completion of hyperbolic groups, their universal theory and discrimination* (in preparation).
12. G. Baumslag, A. G. Myasnikov, and V. Romankov, *Two theorems about equationally Noetherian groups*, J. Algebra **194** (1977), 654–664.
13. G. Baumslag and D. Solitar, *Some two-generator one-relator non-Hopfian groups*, Bull. Amer. Math. Soc. **68** (1962), 199–201.

14. D. E. Cohen and R. C. Lyndon, *Free bases for normal subgroups of free groups*, Trans. Amer. Math. Soc. **108** (1963), 526–537.
15. D. B. A. Epstein, J. W. Cannon, D. F. Holt, S. V. F. Levy, M. S. Paterson, and W. P. Thurston, *Word processing in groups*, Jones and Bartlett, Boston, 1992.
16. S. M. Gersten, *Dehn functions and l_1-norms of finite presentations*, Algorithms and Classification in Combinatorial Group Theory (Berkeley, 1989) (G. Baumslag and C. F. Miller, III, eds.), Math. Sci. Res. Inst. Publ., vol. 23, Springer, New York, 1992, pp. 195–224.
17. É. Ghys, A. Haefliger, and A. Verjovsky (eds.), *Group theory form a geometrical viewpoint* (Trieste, 1990), World Scientific, River Edge, 1991.
18. M. Gromov, *Hyperbolic groups*, Essays in Group Theory (S. M. Gersten, ed.), Math. Sci. Res. Inst. Publ., vol. 8, Springer, New York-Berlin, 1987, pp. 75–263.
19. V. S. Guba, *Equivalence of infinite systems of equations in free groups and semigroups to finite subsystems*, Mat. Zametki **40** (1986), no. 3, 321–324 (Russian); English transl., Math. Notes **40** (1986), no. 3-4, 688–690.
20. A. Juhash, *Conjugacy problem for one-relator groups* (unpublished).
21. A. Karrass and D. Solitar, *One-relator groups are virtually torsion-free* (unpublished).
22. R. C. Lyndon, *Groups with parametric exponents*, Trans. Amer. Math. Soc. **96** (1960), 518–533.
23. R. C. Lyndon and P. E. Schupp, *Combinatorial group theory*, Ergeb. Math. Grenzgeb., vol. 89, Springer-Verlag, Berlin-New York, 1977.
24. W. Magnus, *Ueber diskontinuierliche Gruppen mit einer definierenden Relation (Der Frehietssatz)*, J. Reine Angew. Math. **163** (1930), 141–165.
25. ———, *Das Identitatsproblem fuer Gruppen mit einer definierenden Relation*, Math. Ann. **106** (1932), 295–307.
26. W. Magnus, A. Karrass, and D. Solitar, *Combinatorial group theory: presentations of groups in terms of generators and relations*, Interscience, New York-London-Sydney, 1966.
27. ———, *Elements of finite order in groups with a single defining relation*, Comm. Pure Appl. Math **13** (1960), 57–66.
28. A. I. Malcev, *On isomorphic representations of infinite groups*, Mat. Sbornik **8(50)** (1940), 405–422 (Russian).
29. A. G. Myasnikov and V. N. Remeslennikov, *Degree groups. I. Foundations of the theory and tensor completions*, Sibirsk. Mat. Zh. **35** (1994), no. 5, 1106–1118 (Russian); English transl., Siberian Math. J. **35** (1994), no. 5, 986–996.
30. ———, *Exponential groups. II. Extension of centralizers and tensor completion of CSA-groups*, Internat. J. Algebra Comput. **6** (1996), no. 6, 687–711.
31. *Identical relations in groups.* I, Math. Ann. **114** (1937), 506–537.
32. H. Neumann, *Varieties of groups*, Springer-Verlag, New York, 1967.
33. B. B. Newman, *Some results in one-relator groups*, Bull. Amer. Math. Soc. **74** (1968), 568–571.
34. V. N. Remeslennikov, *∃-free groups*, Siberian Math. J. **30** (1989), no. 6, 153–157 (Russian); English transl., Siberian Math. J. **30** (1989), no. 6, 998–1001.
35. Z. Sela, *The isomorphism problem for hyperbolic groups*, Ph.D. thesis, Hebrew University of Jerusalem, 1991.
36. ———, *The isomorphism problem for hyperbolic groups.* I, Ann. of Math. (2) **141** (1995), no. 2, 217–283.
37. J. R. Stallings, *Finiteness of matrix representations*, Ann. of Math. (2) **124** (1986), no. 2, 337–346.

Department of Mathematics, City College of New York, City University of New York, New York, NY 10031, USA

E-mail address: gilbert@groups.sci.ccny.cuny.edu

Centre de Recherches Mathématiques
CRM Proceedings and Lecture Notes
Volume **17**, 1999

Length Functions and Λ-Trees

I. M. Chiswell

1. Introduction

An ordinary tree defines an integer-valued metric on its set of vertices, called the path metric, just by taking the shortest path between two vertices, and defining their distance apart to be the number of edges in the path. The idea of a Λ-tree (Λ being an ordered abelian group) is obtained by considering Λ-valued metrics having certain properties in common with the path metric. Some aspects of the theory of group actions on ordinary trees can be generalized to group actions (as isometries) on Λ-trees. These include a classification of single isometries of Λ-trees, and a construction of a Λ-tree arising from a valuation on a field F, on which the group $\mathrm{GL}_2(F)$ acts. However, there is at present no good analogue of the Bass-Serre structure theory for groups acting on ordinary trees, which appears in [**47**, Chapter I]. Originally it was intended to mention this theory briefly, but eventually the whole of the first lecture was used to discuss it.

The theory of Λ-trees has its origins in two papers from 1976/77, by J. Tits [**54**] and the author [**11**]. Tits' paper contains a definition of an $\mathbb{R}$-tree, although the requirement that it should be a complete metric space is not needed for most purposes, and in recent work is assumed only when necessary. The author's paper contains a construction of an $\mathbb{R}$-tree starting from a Lyndon length function on a group, an idea considered earlier by Lyndon [**35**], who introduced integer-valued length functions to generalize the Nielsen method in free groups and free products. Real-valued length functions were considered by Harrison [**31**], whose results will be mentioned later. However, the author did not show that the space constructed in [**11**] is an $\mathbb{R}$-tree. This was left to Imrich [**32**] and (in perhaps a more accessible reference), Alperin and Moss [**2**]. It should be noted that these authors used various different definitions of an $\mathbb{R}$-tree, which we shall discuss later.

The theory of $\mathbb{R}$-trees came into prominence with the work of Morgan and Shalen [**39**]. They showed that if G is a finitely generated group, then the space of discrete, faithful representations of G as orientation-preserving isometries of hyperbolic space $\mathbb{H}^n$, factored out by conjugacy in the group $\mathrm{SO}(n,1)$, has a compactification. The ideal points are obtained from certain actions of G on $\mathbb{R}$-trees. This paper also contains the definition of Λ-tree, and develops the theory of Λ-trees for

1991 *Mathematics Subject Classification.* 20E08.
This is the final form of the paper.

arbitrary ordered abelian groups Λ. In particular, the construction in the author's paper [**11**] mentioned above generalises to arbitrary Λ. The present version of this result is given in Theorem 3.4 below, and a version for Λ-valued Lyndon length functions is proved as a corollary in Theorem 3.5. The work of Morgan and Shalen is described in a survey article by Morgan [**38**].

Morgan and Shalen were led to consider Λ-trees by the example of a Λ-tree arising from a valuation on a field, a generalisation of the construction of a tree from a discrete valuation given in [**47**, Chapter II]. In this construction, which we shall briefly describe later, Λ is the value group.

The theory was further developed in an important paper by Alperin and Bass [**1**], which is the standard reference. Another useful paper for the general theory is that by Culler and Morgan [**20**], which confines attention to $\mathbb{R}$-trees. However, the most interesting results in this paper concern spaces of actions on $\mathbb{R}$-trees, motivated by the fact that the compactification by Morgan and Shalen mentioned previously involves points in this space. They show that, if G is a finitely generated group, then the space PLF(G) of projectivised non-zero hyperbolic length functions for actions of G on $\mathbb{R}$-trees is itself compact. We shall have to refer to [**20**] for the exact definition of this space and the proof, but hyperbolic length functions will be discussed later for arbitrary Λ-trees. They also show that the space SLF(G) of points in PLF(G) arising from what are called small actions of G on $\mathbb{R}$-trees is a compact subspace. An action is called *small* if the stabilizer of every segment (as defined below) contains no free subgroup of rank 2. This result is of interest because the points in the Morgan-Shalen compactification belong to SLF(G).

There have been some interesting recent developments concerning spaces of actions, which we shall not be able to discuss in detail. Let F_n be the free group of rank n, Out(F_n) its group of outer automorphisms. Then Outer Space, the space introduced by Culler and Vogtmann [**21**] on which Out(F_n) acts, may be viewed as the subspace of SLF(F_n) consisting of points arising from free actions of F_n on polyhedral $\mathbb{R}$-trees, as defined below. It has been shown by Bestvina and Feighn [**8**], building on work of Cohen and Lustig [**18**], that the closure of Outer Space in SLF(F_n) consists of the points arising from what are called very small actions on $\mathbb{R}$-trees. They showed that for $n \geq 2$, this closure has dimension $3n-4$, and it was shown by Gaboriau and Levitt [**24**] that the boundary of outer space has dimension $3n-5$. In the case $n = 2$, Outer Space has been explicitly described in [**22**]. Culler and Vogtmann [**21**] showed that Outer Space is contractible, and there has been further work on the contractibility of related spaces, including its closure. See [**33**], [**51**], [**53**] and [**58**]. For $n = 2$, any free action of F_2 on an $\mathbb{R}$-tree which is minimal, in that there is no non-empty proper F_2-invariant subtree, must be an action on a polyhedral $\mathbb{R}$-tree. This follows from Proposition 4.2 in [**22**] and the argument for Theorem 1 in [**13**]. (This argument shows that the "division process" in [**22**] must terminate, because the group M in the proof is archimedean). However, there are free, minimal actions of F_n for $n > 2$ which are not polyhedral. Examples have been given by Bestvina and Feighn (see §2.2 in [**49**]) and by Levitt [**34**].

We begin with the Bass-Serre theory, reflecting the contents of the first lecture. The rest of the material, on Λ-trees, is an introduction to the theory, and is based on lectures given at an earlier conference at the University of York, England, which were written up in an expanded form in [**14**]. Essentially, §3 and §4 are an abbreviated version of [**14**], but with some updates and several minor additions.

For further reading, we recommend the survey articles by Shalen ([**48**] and [**49**]), the second of which gives a remarkable overview of the subject.

2. The Bass-Serre Theory

The Bass-Serre theory gives a structure theorem for a group acting on an ordinary tree. It is described as the "fundamental group of a graph of groups". This means it can be built up using the basic constructions of combinatorial group theory (free products with amalgamation and HNN-extensions) using direct limits. It was one of the major breakthroughs in combinatorial group theory in the 1960's. Despite the apparent complication of its details, it provided a conceptually simple way of proving subgroup theorems (if a group acts on a tree, so does any subgroup). For example, the Nielsen-Schreier Theorem (a subgroup of a free group is a free group) and the Kurosh Subgroup Theorem (a subgroup of a free product G has a free product decomposition related to that of G) are easy consequences. It also provides information on subgroups of free products with amalgamation and HNN-extensions; such results had hitherto been proved by complicated arguments involving generalized versions of the idea of a Schreier transversal for a subgroup.

The idea of a graph is a familiar one, used in many articles in this volume, but there are various ways of formally defining it. In this section we view them as combinatorial objects. The idea of a directed graph is standard. A directed graph X consists of two sets, $V(X)$ and $E(X)$ (the set of *vertices* and *edges* of X, respectively), together with two mappings $o,\ t\colon E(X) \to V(X)$. We require also that $V(X) \cap E(X) = \emptyset$. The vertices $o(e)$, $t(e)$ (the *origin* and *terminus* of e respectively) are called the endpoints of e, and as usual we depict vertices as points, and edges as line segments joining their endpoints, with an arrow pointing from the origin to the terminus.

One thinks of the edges as routes between certain locations (represented by the vertices), and we would like to be able to traverse an edge in either direction (from $o(e)$ to $t(e)$ or the other way round). For this and other reasons we define a *graph* to be a directed graph with involution, that is, a directed graph X together with a mapping $E(X) \to E(X)$, $e \mapsto \bar{e}$, satisfying: $e \neq \bar{e}$, $\bar{\bar{e}} = e$, $o(\bar{e}) = t(e)$ and (consequently) $t(\bar{e}) = o(e)$, for all $e \in E(X)$. We define an unoriented edge of X to be a pair $\{e, \bar{e}\}$, where $e \in E(X)$. These are depicted as line segments without an arrow.

We can now define a path in a graph X to be a sequence of edges $e_1, \ldots, e_n$ of X with $t(e_i) = o(e_{i+1})$ for $1 \leq i \leq n-1$. We put $o(p) = o(e_1)$, $t(p) = t(e_n)$, and call n the length of the path. In addition, for every vertex v of X, we introduce by convention a *trivial path* 1_v with $o(1_v) = t(1_v) = v$, of length 0. A graph X is called *connected* if, given x, $y \in V(X)$, there exists a path p in X with $o(p) = x$ and $t(p) = y$. This is a combinatorial analogue of a path-connected topological space.

A path $e_1, \ldots, e_n$ is called *reduced* if $e_i \neq \bar{e}_{i+1}$ for $1 \leq i \leq n-1$ and a path is called *closed* if its endpoints coincide (paths of length 0 are regarded as reduced). A *circuit* is a closed reduced path of positive length, $e_1, \ldots, e_n$ such that $o(e_i) \neq o(e_j)$ for $1 \leq i, j \leq n$ with $i \neq j$. Thus a circuit of length 1 consists of a single edge e with $o(e) = t(e)$, a circuit of length 2 consists of two edges e, f with $o(e) = t(f)$, $t(e) = o(f)$ but $e \neq \bar{f}$. For $n > 2$, a circuit of length n, when drawn in the usual manner, looks like a polygon with n sides. A circuit can be viewed as the combinatorial analogue of a homeomorphic image of a circle in a topological space.

A tree is a connected graph without circuits. We shall sometimes refer to a tree in this sense as a simplicial tree, to distinguish it from the notion of a Λ-tree which will be introduced later. This differs from the usage in some other papers in this volume, where the term simplicial tree means the $\mathbb{R}$-tree arising from a tree in the sense we have defined. This will be discussed later.

A morphism of graphs (or graph mapping) from X to Y is a mapping $f: V(X) \cup E(X) \to V(Y) \cup E(Y)$ which maps vertices to vertices and edges to edges, such that, for all edges $e \in V(X)$, $f(o(e)) = o(f(e))$, $f(t(e)) = t(f(e))$ and $f(\bar{e}) = \overline{f(e)}$. A graph X is a subgraph of a graph Y if $V(X) \subseteq V(Y)$, $E(X) \subseteq E(Y)$ and the inclusion map $V(X) \cup E(X) \hookrightarrow V(Y) \cup E(Y)$ is a graph mapping, that is, if $e \in E(X)$ then $o(e)$, $t(e)$ and $\bar{e}$ have the same meaning in Y as they do in X. The graph mapping f is called an isomorphism if it is bijective, and an isomorphism from X to X is called an automorphism of X.

We now have a natural notion of (left) action of a group G on a graph X. The set $\mathrm{Aut}(X)$ of automorphisms of X is a group under composition of mappings, and an action of G on X is given by a homomorphism $G \to \mathrm{Aut}(X)$. It consists of actions on both $V(X)$ and $E(X)$, such that $go(e) = o(ge)$, $gt(e) = t(ge)$ and $g\bar{e} = \overline{ge}$ for all $g \in G$ and $e \in E(X)$.

DEFINITION. An action of a group G on a graph X is *without inversions* if $ge \neq \bar{e}$ for all $g \in G$ and $e \in E(X)$.

When the action is without inversions, we can form the quotient graph $Y = X/G$ as follows. We define $V(Y)$ to be the set of G-orbits of $V(X)$ and $E(Y)$ to be the set of G-orbits of $E(X)$. We put $o(Ge) = Go(e)$, $t(Ge) = Gt(e)$ and $\overline{Ge} = G\bar{e}$. The condition that the action is without inversions ensures $\overline{Ge} \neq Ge$, and Y is a graph. There is a mapping $p: X \to Y$ given by $p(z) = Gz$ for $z \in V(X) \cup E(X)$, and it is a graph map. We call p the quotient map.

We now come to one of the main ideas in the Bass-Serre theory.

DEFINITION. A graph of groups $(\mathcal{G}, Y)$ consists of the following.

(1) A connected graph Y.
(2) For all $v \in V(Y)$, a group G_v.
(3) For all $e \in E(Y)$, a group G_e, with $G_e = G_{\bar{e}}$.
(4) For all $e \in E(Y)$, a monomorphism $i_e: G_e \to G_{o(e)}$.

Given a graph of groups, we shall associate a group with it, called its fundamental group. It will be described by a presentation (group presentations are discussed in the article by G. Baumslag in this volume). If G is any group, the standard presentation $\langle G \mid \mathrm{rel}(G)\rangle$ is obtained by taking a set X in one-to-one correspondence with G, via a mapping $g \mapsto x_g$. The standard presentation is then $\langle X \mid R\rangle$, where R is the set of relations $\{x_g x_h = x_{gh} \mid g, h \in G\}$. Of course, there are many choices for the set X, for example we could take $X = G$ and $x_g = g$ for $g \in G$. Doing this, however, would require some notational care in describing the relations. On the other hand, when dealing with a family of groups $\{G_i \mid i \in I\}$, we can assume that the generating sets in the presentations $\langle G_i \mid \mathrm{rel}(G_i)\rangle$ are pairwise disjoint. This will be done in the definition which follows.

Let $(\mathcal{G}, Y)$ be a graph of groups, and let T be a maximal tree of Y. That is, T is a subtree of Y (i.e. a subgraph which is a tree) with $V(T) = V(Y)$. For the existence of maximal trees, see [**17**], Chapter 5, Proposition 7 et seq. or [**47**], Chapter 1, §2.3, Proposition 11.

DEFINITION. The fundamental group $\pi(\mathcal{G}, Y, T)$ is the group with presentation

$$\langle t_e(e \in E(Y)),\ G_v(v \in V(Y)) \mid \mathrm{rel}(G_v), t_e i_e(g) t_e^{-1} = i_{\bar{e}}(g)\ (g \in G_e, e \in E(Y)),$$
$$t_e t_{\bar{e}} = 1\ (e \in E(Y)),\ t_e = 1\ (e \in E(T))\rangle.$$

Thus the set of generators is the disjoint union of a collection consisting of a set $\{t_e \mid e \in E(Y)\}$ in one-to-one correspondence with $E(Y)$ and a set in one-to-one correspondence with G_v, for each $v \in V(Y)$. The set of relations is the union of $\mathrm{rel}(G_v)$ for each $v \in V(Y)$ and the three sets $\{t_e i_e(g) t_e^{-1} = i_{\bar{e}}(g) \mid g \in G_e, e \in E(Y)\}$, $\{t_e t_{\bar{e}} = 1 \mid e \in E(Y)\}$ and $\{t_e = 1 \mid e \in E(T)\}$. Of course, in the first of these three sets, $i_e(g)$ means the generator in the standard presentation $\langle G_{o(e)} \mid \mathrm{rel}(G_{o(e)})\rangle$ corresponding to $i_e(g)$, and similarly for $i_{\bar{e}}(g)$. This presentation looks complicated, and to help understand it, we first consider two simple cases, where there is only one unoriented edge.

EXAMPLE 1. Let $E(Y) = \{e, \bar{e}\}$, $V(Y) = \{o(e), t(e)\}$ and suppose $o(e) \neq t(e)$. Let $A = G_{o(e)}$, $B = G_{t(e)}$, $C = G_e = G_{\bar{e}}$. Then the only choice of maximal tree is $T = Y$, so we obtain the presentation

$$\langle t_e, t_{\bar{e}}, A, B \mid \mathrm{rel}(A), \mathrm{rel}(B),\ t_e i_e(c) t_e^{-1} = i_{\bar{e}}(c),\ t_{\bar{e}} i_{\bar{e}}(c) t_{\bar{e}}^{-1} = i_e(c)\ (c \in C),$$
$$t_e t_{\bar{e}} = 1,\ t_{\bar{e}} t_e = 1,\ t_e = 1, t_{\bar{e}} = 1\rangle.$$

After applying Tietze transformations, we obtain

$$\langle A, B \mid \mathrm{rel}(A), \mathrm{rel}(B),\ i_e(c) = i_{\bar{e}}(c)\ (c \in C)\rangle$$

which is a presentation of the free product with amalgamation $A *_C B$.

EXAMPLE 2. Again let $E(Y) = \{e, \bar{e}\}$, but now suppose $o(e) = t(e)$ and $V(Y) = \{o(e)\}$. Let $A = G_{o(e)}$ and $C = G_e = G_{\bar{e}}$. Again there is only one choice of maximal tree T, namely $V(T) = V(Y)$ and $E(T) = \emptyset$. Then after Tietze transformations we obtain the presentation

$$\langle t, A \mid \mathrm{rel}(A), t i_e(c) t^{-1} = i_{\bar{e}}(c)\ (c \in C)\rangle$$

where $t = t_e$. The group with this presentation is called an HNN-extension of A with stable letter t and associated subgroups $i_e(C)$, $i_{\bar{e}}(C)$.

We can generalize Example 2 by considering a graph of groups where the graph has one vertex, with the group A associated to it, but with any number of edges. The fundamental group has a presentation of a similar form, but instead of a single stable letter t there will be one for each unoriented edge, with a corresponding set of relations for each such edge. The group with this presentation is again called an HNN-extension with base A. For any graph of groups $(\mathcal{G}, Y)$, the group $\pi(\mathcal{G}, Y, T)$ is independent, up to isomorphism, of the choice of maximal tree T. See [**17**], Chapter 8, Proposition 18 and the remarks following it, or [**47**], Chapter I, §5, Proposition 20.

If $(\mathcal{G}, Y)$ is a graph of groups and Z is a connected subgraph of Y, we obtain a graph of groups $(\mathcal{G}|_Z, Z)$, where the vertex and edge groups and monomorphisms are the same as in Y, but the underlying graph is Z. If T is a maximal tree of Y and S is a maximal tree of Z such that S is a subtree of T, there is a natural map $\pi(\mathcal{G}|_Z, Z, S) \to \pi(\mathcal{G}, Y, T)$ induced by the mapping $t_e \mapsto t_e$ for $e \in E(Z)$ and the identity mappings $G_v \to G_v$ for $v \in V(Z)$. It is shown in Lemma 19, Chapter 8 of **[17]** that this map is a monomorphism. In particular, taking Z to consist of a single vertex v with no edges, the natural map $G_v \to \pi(\mathcal{G}, Y, T)$ is an embedding.

The argument in **[17]** shows that if Y is a tree, then $\pi(\mathcal{G}, Y, Y)$ is the direct limit of $\pi(\mathcal{G}|T, T, T)$ taken over all finite subtrees T of Y. Further, if Y is a finite tree, $\pi(\mathcal{G}, Y, Y)$ is built up by successively taking free products with amalgamation. Finally, for any graph of groups $(\mathcal{G}, Y)$, with maximal tree T, $\pi(\mathcal{G}, Y, T)$ is an HNN-extension with base $\pi(\mathcal{G}|T, T, T)$, and stable letters corresponding to unoriented edges of $E(Y)\backslash E(T)$.

Now suppose G is a group acting without inversions on a connected graph X, let $Y = X/G$ be the quotient graph, and let $p: X \to Y$ be the projection map. We shall make Y into a graph of groups, where the groups associated to the vertices and edges of Y are stabilizers of vertices and edges of X for the action of G. Choose a maximal tree T of Y. It is possible to find a graph map $j: T \to X$ such that $pj = \mathrm{id}_T$. (See **[17]** Chapter 8, Lemma 2 and remark after Lemma 5, or **[47]** Chapter I, §3, Proposition 14). Thus $j(T)$ is an isomorphic copy of T.

We want to extend j to all of Y, so that $pj = \mathrm{id}_Y$, but it will be necessary to abandon the requirement that the extension of j is a graph map. We need to define $j(e)$ for $e \in E(Y)\backslash E(T)$. Now such an edge e is $G\widetilde{e}$ for some $\widetilde{e} \in E(X)$. Further, $p(o(\widetilde{e})) = o(e) = pj(o(e))$, so $o(\widetilde{e})$ and $j(o(e))$ are in the same orbit, and we take $g \in G$ such that $go(\widetilde{e}) = j(o(e))$. Then $g\widetilde{e}$ satisfies $o(g\widetilde{e}) = jo(e)$. We could then set $j(e) = g\widetilde{e}$. This would ensure that $o(je) = jo(e)$ for all $e \in E(Y)$. However, to define a graph of groups, it is desirable that j should continue to satisfy $\overline{j(e)} = j(\bar{e})$ for all $e \in E(Y)$.

To achieve this, we take an orientation A of Y. This means that A is a subset of $E(Y)$, containing exactly one element of every pair $\{e, \bar{e}\}$, for $e \in E(Y)$. We then define $j(e) = g\widetilde{e}$ as above, but only for $e \in A\backslash E(T)$, and define $j(\bar{e}) = \overline{j(e)}$. We have now defined j on all of Y in such a way that $pj = \mathrm{id}_Y$ and $\overline{j(e)} = j(\bar{e})$ for all $e \in E(Y)$.

If $e \in A\backslash E(T)$, it is not necessarily true that $jt(e) = t(j(e))$, but $p(j(t(e))) = t(e)$, and $p(t(je)) = t(pj(e)) = t(e)$, so $jt(e)$ and $t(je)$ are in the same orbit, and we choose $\gamma_e \in G$ such that $\gamma_e jt(e) = t(je)$. We also put $\gamma_{\bar{e}} = \gamma_e^{-1}$, and $\gamma_e = 1$ for $e \in E(T)$.

We can now make a graph of groups $(\mathcal{G}, Y)$ by defining $G_z = \mathrm{stab}(j(z))$ for $z \in V(Y) \cup E(Y)$. The monomorphism i_e is the inclusion map if $e \in E(T)$ or $e \in A\backslash E(T)$, and otherwise it is given by $i_e(g) = \gamma_e^{-1} g \gamma_e$. One can check, on unravelling all this that the following are true

$$\begin{aligned} \gamma_e i_e(g) \gamma_e^{-1} &= i_{\bar{e}}(g) && \text{for all } e \in E(Y) \\ \gamma_e \gamma_{\bar{e}} &= 1 && \text{for all } e \in E(Y) \\ \gamma_e &= 1 && \text{for all } e \in E(T). \end{aligned}$$

This means, by the universal property of a presentation, that there is a group homomorphism $\phi: \pi(\mathcal{G}, Y, T) \to G$ induced by the identity mapping on each G_v, for $v \in V(Y)$, and by sending t_e to γ_e. The Bass-Serre Structure Theorem, which we shall not prove, is the following.

THEOREM 2.1. *The homomorphism ϕ just defined is an isomorphism if and only if X is a tree.*

Conversely, given a graph of groups $(\mathcal{G}, Y)$, and a maximal tree T of Y, let $G = \pi(\mathcal{G}, Y, T)$. It is possible to construct a graph $\widetilde{Y}$ on which G acts, such that the quotient $\widetilde{Y}/G$ is isomorphic to Y, and the groups associated to the vertices and edges of Y are stabilizers of vertices and edges of $\widetilde{Y}$. The basic theory of group actions tells us what the edges and vertices of $\widetilde{Y}$ must be, up to G-equivariant isomorphism. We let

$$V(\widetilde{Y}) = \coprod_{v \in V(Y)} G/G_v \quad \text{and} \quad E(\widetilde{Y}) = \coprod_{e \in E(Y)} G/G_e.$$

To finish the definition, we again need to choose an orientation, say A, of Y. Then for $e \in A$ and $g \in G$, we define

$$\begin{aligned} o(gG_e) &= gG_{o(e)} \\ t(gG_e) &= gt_eG_{t(e)} \\ \overline{gG_e} &= gG_{\bar{e}}. \end{aligned}$$

This is sufficient information to define the graph $\widetilde{Y}$. Note that, in an attempt to keep the notation simple, gG_e has been identified with its image in $E(\widetilde{Y})$. However, this can cause problems because, for example, $G_e = G_{\bar{e}}$. With the usual construction of a disjoint union, we should take $E(\widetilde{Y}) = \bigcup_{e \in E(Y)} (G/G_e) \times \{e\}$, and so write (gG_e, e) rather than just gG_e. Similar remarks apply, of course, to $V(\widetilde{Y})$.

Now G acts on G/G_v and G/G_e by left multiplication, and this gives an action of G on $\widetilde{Y}$ as automorphisms, which is clearly without inversions.

THEOREM 2.2. *The graph $\widetilde{Y}$ just constructed is a tree.*

Theorem 2.2 is, in fact, the main point of the Bass-Serre theory, and is the main ingredient in the proof of Theorem 2.1. There are various ways of proving it. See for example, Chapter 8 in [**17**], which uses an argument of the author [**12**]. Together, 2.1 and 2.2 provide a method of proving subgroup theorems. As an illustration, we shall prove two classical theorems.

An action of a group on a tree (or any graph) X is called *free* if $\operatorname{stab}(z) = 1$ for all $z \in V(X) \cup E(X)$.

COROLLARY 2.3. *A group G is a free group if and only if it acts freely and without inversions on a tree.*

PROOF. Suppose G is free with basis S. Let Y be a graph with one vertex and with $E(Y) = S \cup S^{-1}$, where $\bar{s} = s^{-1}$ for $s \in E(Y)$. Let T be the maximal tree of Y (so $V(T) = V(Y)$ and $E(T) = \emptyset$). Assign the trivial group to all edges and vertices of Y, to obtain a graph of groups $(\mathcal{G}, Y)$. The fundamental group $\pi(\mathcal{G}, Y, T)$ has (after Tietze transformations) the presentation

$$\langle t_e(e \in E(Y)) \mid t_e t_{\bar{e}} = 1(e \in E(Y)) \rangle$$

and after further Tietze transformations we obtain $\langle t_s(s \in S) \rangle$, with the empty set of relations. This is a free group with basis $\{t_s \mid s \in S\}$, so is isomorphic to G. Therefore G acts without inversions on the tree $\widetilde{Y}$ in Theorem 2.2, and the action is free since all edge and vertex groups of $(\mathcal{G}, Y)$ are trivial.

Conversely if G acts freely and without inversions on a tree X, by Theorem 2.1 G is isomorphic to the fundamental group of a graph of groups $\pi(\mathcal{G}, Y, T)$, where all the edge and vertex groups of $(\mathcal{G}, Y)$ are trivial. It follows by a similar argument to that in the first part that $\pi(\mathcal{G}, Y, T)$ is a free group, with basis $\{t_e \mid e \in A \backslash E(T)\}$, where A is an orientation of Y.

Let us look more closely at the tree $\widetilde{Y}$ in the first part of the proof of 2.3. We identify G with $\pi(\mathcal{G}, Y, T)$, so t_s is identified with s. Since Y has one vertex and all edge and vertex groups are trivial, we may take $V(\widetilde{Y}) = G$ and $E(\widetilde{Y}) = \bigcup_{e \in S \cup S^{-1}} G \times \{e\}$, that is, $E(\widetilde{Y}) = G \times (S \cup S^{-1})$. Taking S as the orientation of Y, for $s \in S$ we have $o(g,s) = g$, $t(g,s) = gs$ and $\overline{(g,s)} = (g, s^{-1})$. This is one description of the Cayley graph of G with respect to S. (There are alternative descriptions in other articles in this volume). One can prove directly, using the normal form theorem for free groups, and the fact that S generates G, that $\widetilde{Y}$ is a tree, without recourse to Theorem 2.2.

COROLLARY 2.4 (Nielsen-Schreier Theorem). *A subgroup of a free group is free.*

PROOF. Let G be a free group, H a subgroup of G. By 2.3, G acts freely and without inversions on a tree, hence so does H by restriction, and again by 2.3 H is a free group.

There are easier ways of proving 2.4 than by first proving Theorems 2.1 and 2.2 in complete generality. However, it is worth noting that 2.3 is the simplest special case of the general theory. There are many other applications of the theory, including the next corollary.

COROLLARY 2.5 (Kurosh Subgroup Theorem). *If $G = \ast_{i \in I} G_i$ is a free product, and H is a subgroup of G, then H has a decomposition*

$$H = F * \mathop{\ast}_{i \in I} \mathop{\ast}_{g \in T_i} H \cap gG_ig^{-1}$$

where F is a free group and and T_i is some suitable set of representatives for the double cosets $\{HgG_i \mid g \in G\}$.

PROOF. Let Y be the graph with $V(Y) = I \cup \{0\}$ (0 being some element not in I) and $E(Y) = \big(I \times \{0\}\big) \cup \big(\{0\} \times I\big)$, where $o(0,i) = 0$, $t(0,i) = i$ and $\overline{(0,i)} = (i,0)$, etc. Then Y is a tree. Associate the group G_i with the vertex i, and the trivial group to all edges and the vertex 0, to obtain a graph of groups $(\mathcal{G}, Y)$. After Tietze transformations the presentation of $\pi(\mathcal{G}, Y, Y)$ has the form $\big\langle G_i \ (i \in I) \mid \mathrm{rel}(G_i) \ (i \in I)\big\rangle$, so G is isomorphic to $\pi(\mathcal{G}, Y, Y)$. By Theorem 2.2 there is a tree $\widetilde{Y}$ on which G acts without inversions, and H acts by restriction. Note that $V(\widetilde{Y}) = \big(\coprod_{i \in I} G/G_i\big) \coprod G$.

Let $Z = \widetilde{Y}/H$ be the quotient graph, choose a maximal tree T and orientation A of Z and form a graph of groups $(\mathcal{H}, Z)$ for the action of H on $\widetilde{Y}$, using a mapping $j \colon Z \to \widetilde{Y}$ as above. By Theorem 2.1, H is isomorphic to $\pi(\mathcal{H}, Z, T)$. All edge groups of $(\mathcal{H}, Z)$ are trivial, and after Tietze transformations we see that $\pi(\mathcal{H}, Z, T)$ is a free product $F * \ast_{v \in V(Z)} H_v$, where H_v is the group associated to v in $(\mathcal{H}, Z)$, and F is a free group with basis $\{t_e \mid e \in A \backslash E(T)\}$. If $j(v) = gG_i$ for some $i \in I$ (more accurately, $j(v) = (gG_i, i)$), then $H_v = H \cap gG_ig^{-1}$, the stabilizer of $j(v)$ in H. For each $i \in I$, the vertices $j(v)$ of this form constitute a set of representatives for the orbits of H acting on G/G_i, which are in one-to-one correspondence with the double cosets HgG_i ($g \in G$), where the H-orbit of gG_i is associated to HgG_i. The remaining vertex groups of $(\mathcal{H}, Z)$ are trivial, and the result follows.

We leave it as an exercise to describe the tree $\widetilde{Y}$ in the proof of Corollary 2.5, and to prove it is a tree using the normal form theorem for free products. In fact, Theorem 2.2 may be viewed as a kind of normal form theorem for a general graph of groups.

There is a topological interpretation of the Bass-Serre theory. Given a graph of groups $(\mathcal{G}, Y)$, one can construct a complex X whose fundamental group is isomorphic to the fundamental group of $(\mathcal{G}, Y)$. Further, the graph $\widetilde{Y}$ in 2.2 reflects the structure of the universal covering $\widetilde{X}$ of X. The ideas of fundamental group and covering space are discussed in Nadia Benakli's article in this volume. The details of this topological approach are contained in an article by Wall and Scott [**57**] (see also Propositions 22 and 23, Chapter 8 in [**17**]). One can use this method to prove the main theorems of Bass-Serre theory. In the author's view, the approach outlined here is simpler. However, the topological point of view is illuminating, and topologists may find it easier.

It was in realizing there is such a topological interpretation that the author, in his thesis, was able to adapt an argument of Stallings to prove the other main theorem about free products, Grushko's Theorem, using the Bass-Serre theory. In its simplest form, Grushko's Theorem states that, if $G = A * B$ is a free product, then $d(G) = d(A) + d(B)$, where $d(G)$ means the minimum number of generators of G. Over the years, the author's argument has been simplified, and the present-day version can be found in Chapter 8 of [**17**], as Corollary 1 to Theorem 40. The final simplification was made by Stallings himself [**52**], who gave a neat proof that certain "foldings" of G-trees lead to G-trees (a G-tree being a tree with a group G acting on it). Stallings also gives some interesting further applications of this method.

3. Λ-Trees

Let Λ be an ordered abelian group, written additively. This means that there is a linear ordering $\leq$ on Λ such that $a \leq b$ implies $a + c \leq b + c$, for all a, b and $c \in \Lambda$. The following two examples will suffice to illustrate the general theory which follows.

EXAMPLE 1. Any subgroup of the additive group of $\mathbb{R}$ is an ordered abelian group. Such ordered abelian groups Λ are called archimedean; they satisfy the property that if a, $b \in \Lambda$ and $a \neq 0$, then there exists $n \in \mathbb{Z}$ such that $b < na$. In particular, $\mathbb{R}$ itself and $\mathbb{Z}$ are archimedean ordered abelian groups.

EXAMPLE 2. If Λ_1 and Λ_2 are ordered abelian groups, the direct sum $\Lambda_1 \oplus \Lambda_2$ can be made into an ordered abelian group by defining $(a, b) \leq (c, d)$ if and only if either $a < c$, or $a = c$ and $b \leq d$. This is called the *lexicographic ordering*. Clearly the lexicographic ordering extends inductively to the direct sum of finitely many ordered abelian groups. In particular, $\Lambda = \mathbb{Z} \oplus \mathbb{Z}$ is an ordered abelian group with the lexicographic ordering. This is non-archimedean: if $a = (0, 1)$ and $b = (1, 0)$, then $na < b$ for all $n \in \mathbb{Z}$.

Let Λ be an ordered abelian group. It makes sense to speak of a metric with values in Λ. A Λ-valued metric on a set X is a mapping $d: X \times X \to \Lambda$ satisfying the usual axioms ($d(x,y) \geq 0$, $d(x,y) = 0$ if and only if $x = y$, $d(x,y) = d(y,x)$, and the triangle inequality: $d(x,y) \leq d(x,z) + d(z,y)$, for all x, y, $z \in X$). We call a pair (X,d), where d is a Λ-valued metric on X, a Λ-metric space.

EXAMPLE 3. Let Y be a connected graph, let $X = V(X)$, and let d be the path metric on X, that is, $d(x,y)$ is the minimum length of a path in Y with endpoints x, y. Then (X,d) is a $\mathbb{Z}$ metric space.

EXAMPLE 4. There is a Λ-valued metric on Λ itself, for any ordered abelian group Λ, defined by $d(x,y) = |x-y|$, where, for $a \in \Lambda$, $|a| = \max\{a, -a\}$.

We define closed intervals in Λ just as for $\mathbb{R}$, that is, $[a,b]_\Lambda = \{x \in \Lambda \mid a \leq x \leq b\}$ for a, $b \in \Lambda$ with $a \leq b$, and $[b,a]_\Lambda = [a,b]_\Lambda$. We can similarly define open and half open intervals, but these will not be needed.

DEFINITION. A segment in a Λ-metric space (X,d) is the image of an isometry $\alpha: [a,b]_\Lambda \to X$ for some a, $b \in \Lambda$ with $a \leq b$. Note that $a = b$ is allowed. The endpoints of the segment are $\alpha(a)$, $\alpha(b)$.

The endpoints of the segment are characterised as the points of the segment at maximum distance apart. A Λ-metric space (X,d) is called *geodesic* if, for all x, $y \in X$, there is a segment in X with endpoints x, y. In the case $\Lambda = \mathbb{R}$, this accords with the definition in S. Gersten's article in this volume. However, it does depend on Λ, and a geodesic Λ-metric space need not be path connected. Indeed it can be discrete. For it is easy to see that a $\mathbb{Z}$ metric space (X,d) is geodesic if and only if there is a connected graph Y with $X = V(Y)$ and d the path metric on X. (This is left as an exercise; given a geodesic $\mathbb{Z}$ metric space (X,d), Y is obtained by joining two elements x, y of X by an unoriented edge if and only if $d(x,y) = 1$).

This raises the question, given a geodesic $\mathbb{Z}$-metric space (X,d), when is X the set of vertices of a tree, such that d is the path metric? One answer is given by the following, which was used by Morgan and Shalen as a model for their definition of a Λ-tree.

PROPOSITION 3.1. *Let (X,d) be a geodesic $\mathbb{Z}$ metric space. Then there is a tree Γ, such that $X = V(\Gamma)$ and d is the path metric of Γ, if and only if*

(*) *if two segments of (X,d) intersect in a single point, which is an endpoint of both, then their union is a segment.*

PROOF. We refer to [**14**] for the proof of this.

The following is the original definition of a Λ-tree by Morgan and Shalen.

DEFINITION. A Λ-tree is a Λ-metric space (X,d) such that:

(a) (X,d) is geodesic.
(b) Condition (*) in 3.1 above holds.
(c) The intersection of two segments with a common endpoint is also a segment.

It follows easily from Axiom (c) that, if x, y are points of a Λ-tree (X,d), there is a unique segment whose set of endpoints is $\{x,y\}$, and this segment is denoted by $[x,y]$. In Axiom (b), if $[x,y] \cap [x,z] = \{x\}$, then $\sigma = [x,y] \cup [x,z]$ is a segment. It is not difficult to see that y, z are the points at maximum distance apart in σ, so are the endpoints of σ, that is, $\sigma = [y,z]$. Similarly in Axiom (c), one can show that, if

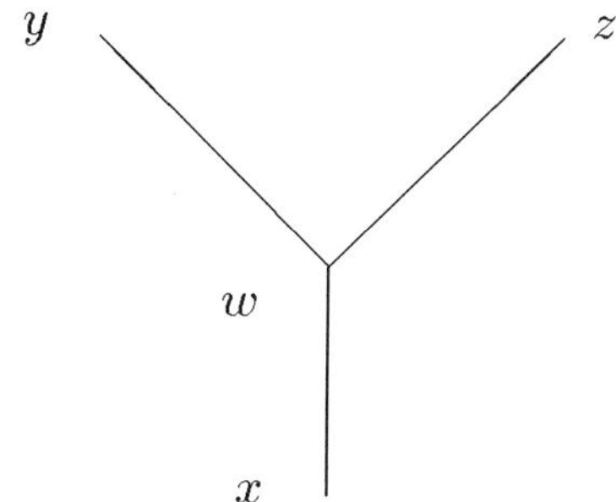

FIGURE 1. A geodesic triangle in a Λ-tree.

x, y, $z \in X$, then x is an endpoint of $[x,y] \cap [x,z]$, so $[x,y] \cap [x,z] = [x,w]$ for some unique $w \in X$. Further, this is symmetric in x, y and z. That is, $[x,y] \cap [y,z] = [y,w]$ and $[x,z] \cap [y,z] = [z,w]$. This follows because $[y,z] = [y,w] \cup [w,z]$ by Axiom (b). Thus the diagram above is a picture of a geodesic triangle in X. It follows that geodesic triangles in a Λ-tree are δ-thin with $\delta = 0$, in the sense of S. Gersten's article in this volume (the definition works for arbitrary ordered abelian groups Λ).

EXERCISE. If x, $v \in X$ (where (X,d) is a Λ-tree) and x_m denotes the point at distance m from v in $[v,x]$, where $0 \le m \le d(v,x)$, show that $d(x_m, y_n) = m + n - 2\min\{m, n, d(v,w)\}$, where $[v,x] \cap [v,y] = [v,w]$.

We now give some examples of Λ-trees.

EXAMPLE 5. The metric space (Λ, d), where d is given by $d(a,b) = |a-b|$ is a Λ-tree. It is easy to see that a segment in Λ must be of the form $[a,b]_\Lambda$ for some a, $b \in \Lambda$.

EXAMPLE 6. A $\mathbb{Z}$-metric space (X,d) is a $\mathbb{Z}$-tree if and only if there is a simplicial tree Γ such that $X = V(\Gamma)$ and d is the path metric of Γ. This follows from Proposition 3.1. (Axiom (c) is automatically satisfied in any $\mathbb{Z}$ metric space).

EXAMPLE 7. Our next example is an $\mathbb{R}$-tree, the geometric realization real (Y) of a simplicial tree Y, in which we view the unoriented edges as part of the tree, each of them being metrically isomorphic to the unit interval $I = [0,1]_\mathbb{R}$, with endpoints identified appropriately. The metric is defined in a similar way to the path metric. For a full discussion of this example, we refer to [**14**], §2, Example 3. A polyhedral $\mathbb{R}$-tree is one homeomorphic to real(Y) for some simplicial tree Y. This is the terminology of Shalen [**49**]; other authors use "simplicial" rather than "polyhedral", but we have used this term to indicate an ordinary tree.

EXAMPLE 8. Let $X = \mathbb{R}^2$ be the plane, but with metric d defined by

$$d((x_1,y_1),(x_2,y_2)) = \begin{cases} |y_1| + |y_2| + |x_1 - x_2| & \text{if } x_1 \neq x_2 \\ |y_1 - y_2| & \text{if } x_1 = x_2. \end{cases}$$

Thus, to measure the distance between two points not on the same vertical line, we take their projections onto the horizontal axis, and add their distances to these projections and the distance between the projections (distance in the usual euclidean sense). If they are on the same vertical line, their distance is the usual

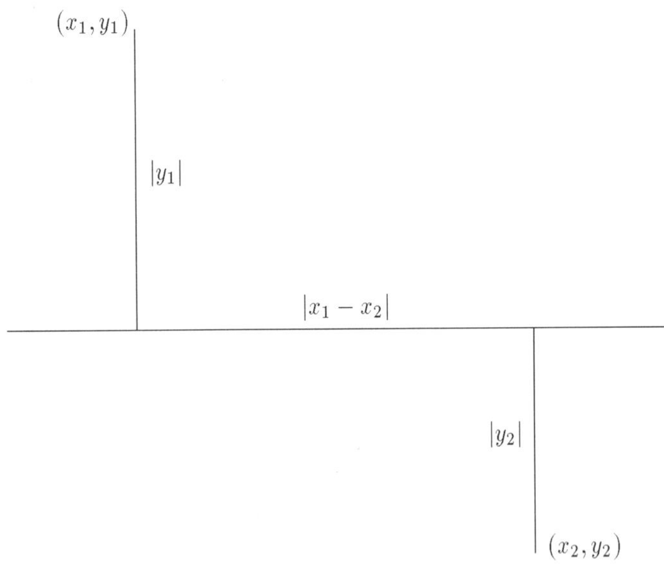

FIGURE 2. Measuring distance in Example 4.

euclidean distance. For a proof that this is an $\mathbb{R}$-tree see [**14**, §2]. This example is illustrated in Figure 2 above.

We shall shortly see that there is a notion of direction at a point in a Λ-tree, and we can therefore define the degree or valency of a point in a Λ-tree. If Y is a simplicial tree, the points of real (Y) of degree greater than 2 are a subset of $V(Y)$ and form a discrete set. However, at each point of the real axis $\mathbb{R} \times \{0\}$ in the present example, there are four directions, and the distance restricted to the real axis is the usual euclidean distance, so the real axis is not a discrete subspace of X. We shall also see that homeomorphisms of $\mathbb{R}$-trees preserve degree of points. It follows that this example is not polyhedral.

Λ-trees as hyperbolic spaces. Suppose (X, d) is a Λ-metric space. Choose a point $v \in X$, and for $x, y \in X$, define

$$(x \cdot y)_v = \tfrac{1}{2}\big(d(x,v) + d(y,v) - d(x,y)\big).$$

This is an element of the ordered abelian group $\frac{1}{2}\Lambda$, which is a subgroup of the ordered abelian group $\mathbb{Q} \otimes_{\mathbb{Z}} \Lambda$. Elements of $\mathbb{Q} \otimes_{\mathbb{Z}} \Lambda$ can be described as fractions a/m, where $a \in \Lambda$, $m \in \mathbb{Z}$ and $m \neq 0$, where $a/m = b/n$ if and only if $na = bm$. Addition is formally the same as the usual addition of fractions, and the ordering is defined by $a/m > 0$ if and only if $ma > 0$. (This is enough to specify the ordering). The group $\frac{1}{2}\Lambda$ is the subgroup $\{a/2 \mid a \in \Lambda\}$.

If (X, d) is a Λ-tree, then $[v, x] \cap [v, y] = [v, w]$ for some w, and $(x \cdot y)_v = d(v, w)$. There is also a simple geometric interpretation of $(x \cdot y)_v$ in real euclidean and hyperbolic space, illustrated in Figure 3 below. Since $d(x, p) = d(x, r)$, $d(y, r) = d(y, q)$ and $d(v, p) = d(v, q)$, it follows easily that $(x \cdot y)_v = d(v, p) = d(v, q)$.

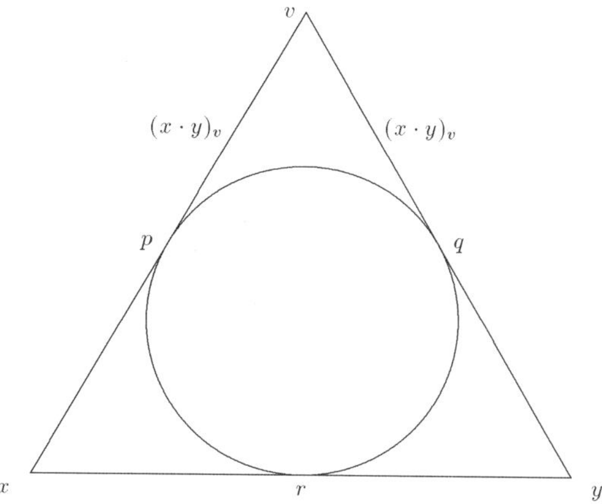

FIGURE 3. Interpretation of $(x \cdot y)_v$ in $\mathbb{R}^n$ and $\mathbb{H}^n$.

DEFINITION (Gromov). Let (X, d) be a Λ-metric space, and let $\delta \in \Lambda$ with $\delta \geq 0$. Then (X, d) is δ-hyperbolic with respect to v if, for all x, y, $z \in X$,

$$x \cdot y \geq \min\{x \cdot z, y \cdot z\} - \delta$$

where $x \cdot y$ means $(x \cdot y)_v$, etc.

This is not the same as the definition in terms of thin triangles given in Gersten's article in this volume, but is equivalent to it, at least for geodesic $\mathbb{R}$-metric spaces, subject to changing the parameter δ by a multiple. For a discussion of this, see [**19**], [**29**] or [**50**]. We have noted that in a Λ-tree, triangles are 0-thin, so it follows that $\mathbb{R}$-trees are 0-hyperbolic in our sense. In fact, one can prove directly that Λ-trees are 0-hyperbolic, for any Λ, and this is part of the proof of Proposition 3.3 below.

If t is another point of X, we have, by direct calculation, the identity

$$(**) \qquad (x \cdot y)_t = d(t, v) + (x \cdot y)_v - (x \cdot t)_v - (y \cdot t)_v$$

for all x, $y \in X$.

LEMMA 3.2. *If (X, d) is δ-hyperbolic with respect to v, and t is any other point of X, then (X, d) is 2δ-hyperbolic with respect to t.*

PROOF. We need to show that $(x \cdot y)_t \geq \min\{(x \cdot z)_t, (y \cdot z)_t\} - 2\delta$, for all x, y, $z \in X$. On adding $(x \cdot t)_v + (y \cdot t)_v + (z \cdot t)_v - d(t, v)$ to both sides and using $(**)$ above, it suffices to show that

$$(x \cdot y)_v + (z \cdot t)_v \geq \min\{(x \cdot z)_v + (y \cdot t)_v, (y \cdot z)_v + (x \cdot t)_v\} - 2\delta$$

for all x, y, z and $t \in X$. Now interchanging x and y if necessary, and also interchanging z and t if appropriate, we can assume that $x \cdot z$ is greater than or equal to $x \cdot t$, $y \cdot z$ and $y \cdot t$, where $x \cdot z$ means $(x \cdot z)_v$, etc. Then by assumption $x \cdot y \geq y \cdot z - \delta$ and $z \cdot t \geq x \cdot t - \delta$. The lemma follows on adding these inequalities.

DEFINITION. (X, d) is δ-hyperbolic if it is δ-hyperbolic with respect to all points of X, and is hyperbolic if it is δ-hyperbolic for some $\delta \geq 0$.

By Lemma 3.2, if (X, d) is 0-hyperbolic with respect to one point, then it is 0-hyperbolic with respect to any other point. Also, (X, d) is 0-hyperbolic with respect to v if and only if, for all x, y and $z \in X$, at least two of $(x \cdot y)_v$, $(y \cdot z)_v$, $(z \cdot x)_v$ are equal, and not greater than the third. This can easily be reinterpreted as follows: (X, d) is 0-hyperbolic if and only if, for all x, y, z and $t \in X$, at least two of $d(x,y)+d(z,t)$, $d(x,z)+d(y,t)$, $d(y,z)+d(x,t)$ are equal, and not less than the third. (This is known as the *four point condition*; a geometric interpretation for Λ-trees is provided by 2.15 in [**1**]). The following result characterizes Λ-trees using the idea of a 0-hyperbolic space.

PROPOSITION 3.3. *Assume (X, d) is a Λ-metric space and v is a point of X. Then (X, d) is a Λ-tree if and only if the following three conditions hold*

(1) *(X, d) is geodesic.*
(2) *(X, d) is 0-hyperbolic with respect to v.*
(3) *For all x, $y \in X$, $(x \cdot y)_v \in \Lambda$.*

PROOF. See [**14**, §3].

We have noted that if condition (2) holds, then (X, d) is 0-hyperbolic with respect to any point of X. It is also true that if condition (3) holds, then $(x \cdot y)_t \in \Lambda$ for all x, y and $t \in X$. This follows from equation $(**)$ above.

We now come to a basic theorem, whose genealogy was discussed in the introduction. If conditions (2) and (3) in Proposition 3.3 are satisfied by the Λ-metric space (X, d), then (X, d) can be embedded in a Λ-tree.

THEOREM 3.4. *Let (X, d) be a Λ-metric space and v a point of X such that*

(1) *for all x, $y \in X$, $(x \cdot y)_v \in \Lambda$;*
(2) *(X, d) is 0-hyperbolic with respect to v.*

Then there exist a Λ-tree (X', d') and an isometry $\phi: X \to X'$ such that, if $\psi: X \to Z$ is any isometry of X into a Λ-tree Z, there is a unique isometry $\mu: X' \to Z$ such that $\mu \circ \phi = \psi$.

OUTLINE OF PROOF. As usual $x \cdot y$ means $(x \cdot y)_v$ for x, $y \in X$. Let

$$Y = \{(x, m) \mid x \in X, m \in \Lambda \text{ and } 0 \leq m \leq d(v, x)\}.$$

Define, for (x, m), $(y, n) \in Y$, $(x, m) \sim (y, n)$ if and only if $x \cdot y \geq m = n$. This is an equivalence relation on Y (it is transitive by assumption (2)). Let $X' = Y/\sim$, and let $\langle x, m \rangle$ denote the equivalence class of (x, m). We define the metric by $d'(\langle x, m\rangle, \langle y, n\rangle) = m + n - 2\min\{m, n, x \cdot y\}$. The mapping $\phi: X \to X'$ is defined by $\phi(x) = \langle x, d(x, v)\rangle$. The details of the proof that this works are given in Theorem 3.1 of [**14**]. The idea of the proof is that $\langle x, m\rangle$ should represent the point on $[v, x]$ (more accurately $[\phi(v), \phi(x)]$) at distance m from v. The definition of the distance is motivated by the exercise preceding Example 5 above. Knowing this, the last part is easy to prove. Given an isometry $\psi: X \to Z$ and $x \in X$, denote by x_m the point on the segment in Z with endpoints $\psi(v)$ and $\psi(x)$, at distance m from $\psi(v)$. Then the mapping $\mu: X' \to Z$ defined by $\mu(\langle x, m\rangle) = x_m$ is an isometry by the exercise, and is the only one with the required properties, because isometries of Λ-trees clearly preserve segments.

REMARK. If $\theta: X \to X$ is an isometry, then in Theorem 3.4 there is a unique isometry $\mu: X' \to X'$ such that $\mu \circ \phi = \phi \circ \theta$ (take $\psi = \phi \circ \theta$ in 3.4). It follows that, if G is a group acting as isometries on X, there is an induced action on X', such that $g\phi(x) = \phi(gx)$ for all $g \in G$ and $x \in X$.

There is another useful way of looking at Theorem 3.4, in terms of "rooted Λ-tree data", in §3 of [**1**].

Lyndon length functions. Let G be a group, Λ an ordered abelian group. A mapping $L: G \to \Lambda$ is called a Lyndon length function if

(1) $L(1) = 0$.
(2) For all $g \in G$, $L(g) = L(g^{-1})$.
(3) For all g, h, $k \in G$, $c(g,h) \geq \min\{c(h,k), c(k,g)\}$, where $c(g,h)$ is defined to be $\frac{1}{2}\big(L(g) + L(h) - L(g^{-1}h)\big)$.
(4) For all g, $h \in G$, $c(g,h) \in \Lambda$.

The last axiom is unnecessary for many purposes, but is convenient here; it can always be achieved simply by doubling the length function. We leave it as an exercise to verify that these axioms imply that for all $g \in G$, $L(g) \geq 0$, and the triangle inequality: for all g, $h \in G$, $L(gh) \leq L(g) + L(h)$.

For example, let G be a group acting as isometries on a Λ-tree (X, d) and let $v \in X$. Then defining $L_v(g) = d(v, gv)$ gives a Lyndon length function on G. By applying the isometry g^{-1} to $[v, gv] \cup [v, hv]$, we see that $c(g,h)$ is $d(v,w)$, where $[v, gv] \cap [v, hv] = [v, w]$, that is, $c(g,h) = (gv \cdot hv)_v$. It follows from the fact that X is 0-hyperbolic that L_v satisfies Axiom (3) above, and it clearly satisfies (1) and (2). Also, $c(g,h) = d(v,w) \in \Lambda$, so (4) is satisfied as well.

A special case is when G is a free group with basis S, acting on its Cayley graph with respect to S, and so on the corresponding $\mathbb{Z}$-tree. (See the remarks after Corollary 2.3). The points of this $\mathbb{Z}$-tree are just the elements of G, and taking v to be the point 1, $L_v(g)$ is the length of the reduced word in $S \cup S^{-1}$ representing g, and $c(g,h)$ is the length of the largest common initial segment of the reduced words representing g and h.

Using the previous theorem we can prove that in fact all Lyndon length functions arise from actions on Λ-trees.

THEOREM 3.5. *Let G be a group and $L: G \to \Lambda$ a Lyndon length function. Then there are a Λ-tree (X', d'), an action of G on X' and a point $v' \in X'$ such that $L = L_{v'}$.*

PROOF. Defining $g \approx h$ if and only if $L(g^{-1}h) = 0$ gives an equivalence relation on G (it is transitive by the triangle inequality). Denote the set of equivalence classes by X, and the equivalence class of g by $\langle g \rangle$. We obtain a Λ-metric space (X, d) by defining $d\big(\langle g \rangle, \langle h \rangle\big) = L(g^{-1}h)$ (the triangle inequality for d follows from that for L). Also, G acts as isometries on X by $g\langle h \rangle = \langle gh \rangle$. Direct computation shows that

$$\big(\langle g \rangle \cdot \langle h \rangle\big)_{\langle 1 \rangle} = c(g,h),$$

so by Axioms (3) and (4), (X, d) satisfies the hypotheses of Theorem 3.4 (with $v = \langle 1 \rangle$), and we obtain a corresponding Λ-tree (X', d') and an isometry $\phi: X \to X'$. We denote the point $\big\langle \langle g \rangle, m \big\rangle$ of X' simply by $\langle g, m \rangle$, and take $v' = \langle 1, 0 \rangle$ as basepoint. Thus from the proof of 3.4, $\phi\big(\langle g \rangle\big) = \big\langle g, L(g) \big\rangle$ for $g \in G$, and $v' = \phi(v)$.

By the remark after Theorem 3.4, the action of G on X induces an action of G on X', such that $g\phi(u) = \phi(gu)$ for all $u \in X$ and $g \in G$. Thus $gv' = \big\langle g, L(g) \big\rangle$ for $g \in G$, and by definition of d' in the proof of 3.4, $d'(v', gv') = L(g)$, as required.

Directions in Λ-trees. As promised in the examples of Λ-trees given above, we show how to define directions at a point in a Λ-tree, but leave it to readers to verify the claims made in Example (8) above.

Let (X, d) be a Λ-tree, and let $v \in X$. Put $D = \{[v,x] \mid x \in X \backslash \{v\}\}$. Define $[v,x] \equiv [v,y]$ to mean that $[v,x] \cap [v,y] \neq \{v\}$. This means that, if $[v,x] \cap [v,y] = [v,w]$ (as in the axioms for a Λ-tree), then $w \neq v$. Since $(x \cdot y)_v = d(v,w)$, this is equivalent to $(x \cdot y)_v > 0$. It follows from the fact that (X,d) is 0-hyperbolic that if $(x \cdot y)_v > 0$ and $(y \cdot z)_v > 0$ then $(x \cdot z)_v > 0$. It then follows easily that $\equiv$ is an equivalence relation on D, and the equivalence classes are called directions at v. We can then define the *degree* of v to be the cardinality of the set $(D/\equiv)$ of directions at v.

There is an alternative way of looking at directions in $\mathbb{R}$-trees, and in order to do this we shall prove some lemmas.

LEMMA 3.6. *If (X,d) is a Λ-tree and $x_0, \ldots, x_n \in X$, then*

$$[x_0, x_n] \subseteq \bigcup_{i=1}^{n} [x_{i-1}, x_i].$$

PROOF. If $n = 2$, $[x_0, x_1] \cap [x_1, x_2] = [x_1, w]$ for some w. By the discussion after the definition of Λ-tree,

$$[x_0, x_2] = [x_0, w] \cup [w, x_2] \subseteq [x_0, x_1] \cup [x_1, x_2].$$

If $n \geq 2$, then $[x_0, x_n] \subseteq [x_0, x_{n-1}] \cup [x_{n-1}, x_n]$ by the case $n = 2$, and the result follows by induction on n.

Lemma 3.6 is part of the "Piecewise Geodesic Proposition" in [**1**], and first appeared as II.1.5 in [**39**]. Many results on Λ-trees consist of showing that pictures one draws for ordinary trees reflect the situation for Λ-trees in general, and the Piecewise Geodesic Proposition is one of the main results needed to do this. Unfortunately, its use can sometimes be tedious, needing the consideration of several cases.

LEMMA 3.7. *Let (X,d) be an $\mathbb{R}$-tree, and let $\alpha\colon [a,b]_{\mathbb{R}} \to X$ be a continuous map. If $x = \alpha(a)$ and $y = \alpha(b)$, then $[x,y] \subseteq \mathrm{Im}(\alpha)$.*

PROOF. Let $A = \mathrm{Im}(\alpha)$. Since A is a closed subset of X (being compact), it is enough to show that every point of $[x,y]$ is within distance ε of A, for all real $\varepsilon > 0$.

Given $\varepsilon > 0$, we can partition the domain of α, say $a = t_0 < \cdots < t_n = b$, so that for $1 \leq i \leq n$, $d\big(\alpha(t_{i-1}), \alpha(t_i)\big) < \varepsilon$. Then all points of $\big[\alpha(t_{i-1}), \alpha(t_i)\big]$ are at distance $< \varepsilon$ from A, for $1 \leq i \leq n$. Finally, $[x,y] \subseteq \bigcup_{i=1}^{n} \big[\alpha(t_{i-1}), \alpha(t_i)\big]$, by Lemma 3.6.

LEMMA 3.8. *Let (X,d) be an $\mathbb{R}$-tree, and suppose $v \in X$. Then for x, $y \in X \backslash \{v\}$, $[v,x] \equiv [v,y]$ if and only if x, y are in the same path component of $X \backslash \{v\}$.*

PROOF. If $v \in [x,y]$, then $[x,v] \cap [v,y] = \{v\}$, so $[v,x] \equiv [v,y]$ implies $[x,y] \subseteq X \backslash \{v\}$, hence x, y are in the same path component of $X \backslash \{v\}$.

Conversely, if $\alpha\colon [a,b]_{\mathbb{R}} \to X\backslash\{v\}$ is a continuous map, with $x = \alpha(a)$, $y = \alpha(b)$, then $[x,y] \subseteq \mathrm{Im}(\alpha)$ by Lemma 3.7, so $v \notin [x,y]$, and $[v,x] \cap [v,y] \neq \{v\}$ by the axioms for a Λ-tree, so $[v,x] \equiv [v,y]$.

Thus the directions at a point v in a $\mathbb{R}$-tree (X,d) are in one-to-one correspondence with the path components of $X\backslash\{v\}$. It follows that homeomorphisms of $\mathbb{R}$-trees preserve degrees of points.

Subtrees. There is a natural notion of subtree of a Λ-tree.

DEFINITION. A subtree of a Λ-tree (X,d) is a subset Y of X such that x, $y \in Y$ implies $[x,y] \subseteq Y$.

A subtree Y is itself a Λ-tree with metric equal to the restriction of d to $Y \times Y$. The intersection of a family of subtrees of a Λ-tree is also a subtree. If S is a subset of a Λ-tree (X,d), we define the *subtree spanned by* S to be the intersection of all subtrees of X containing S.

EXERCISE. If $v \in S$, where S is a subset of a Λ-tree, show that the subtree spanned by S is equal to $\bigcup_{x\in S}[v,x]$. (Use the earlier remarks about geodesic triangles in Λ-trees).

DEFINITION. A subtree Y is called *closed* if it is convex-closed, that is, if the intersection of Y with any segment of X is either empty or a segment of X.

For example, a segment is a closed subtree (see [**1**, Corollary 2.16]). An example of a subtree which is not closed is the interval $(0,1)$ in $\mathbb{R}$. The intersection of any collection of subtrees is a subtree, and the intersection of finitely many closed subtrees is a closed subtree. However, the intersection of an arbitrary family of closed subtrees need not be a closed subtree. For example, take $\Lambda = \mathbb{Q}$, $X = \Lambda$, and Y_r to be the segment joining r and 2, where $r \in \mathbb{Q}$ and $0 \leq r < \sqrt{2}$. Then the intersection of $\bigcap_r Y_r$ with the segment $[0,2]_{\mathbb{Q}}$ is not a segment in our sense. In case $\Lambda = \mathbb{Z}$, $X = V(\Gamma)$ for some simplicial tree Γ, the subtrees of X correspond to subtrees of Γ, and they are all closed.
The main result about closed subtrees is the following lemma, which is part of the "Bridge Proposition" in [**1**] .

LEMMA 3.9. *Let A, B be non-empty closed subtrees of a Λ-tree (X,d) such that $A \cap B = \emptyset$. Then there exist unique points $a \in A$, $b \in B$ such that $[a,b] \cap A = \{a\}$, $[a,b] \cap B = \{b\}$ and if $a_1 \in A$, $b_1 \in B$, then $[a,b] \subseteq [a_1,b_1]$.*

PROOF. See 2.17 in [**1**].

We call $[a,b]$ in 3.9 the *bridge* between A and B. The lemma applies in particular when $B = \{b\}$ is a single point and A is a closed subtree. Also, if A is a closed subtree and $a \in A$, we call $[a,a] = \{a\}$ the bridge between A and a. In the situation of 3.9, we define $d(A,B)$ to be $d(a,b)$. If $A \cap B \neq \emptyset$, we put $d(A,B) = 0$.

A Λ-metric space (X,d) has a topology with basis the open balls $B(x,r) = \{y \in X \mid d(x,y) < r\}$, where $x \in X$ and $r \in \Lambda$, $r > 0$, just as in the case $\Lambda = \mathbb{R}$. Any closed subtree of a Λ-tree is closed in this topology (use 3.9 to show the complement is open), but the converse is false in general, because of the example above, where the intersection of a family of closed subtrees is not a closed subtree. However, the converse is true in the case $\Lambda = \mathbb{R}$, for the intersection of a subtree closed in this topology with a segment is a compact connected subset of the segment, hence a segment.

Characterization of $\mathbb{R}$-trees. We present some alternative ways of defining an $\mathbb{R}$-tree which have been used in the literature. An arc in an $\mathbb{R}$-metric space is defined to be a homeomorphic image of a compact interval (i.e. a segment) in $\mathbb{R}$ with more than one point, and its endpoints are the images of the endpoints of the interval under the homeomorphism. Thus any segment with more than one point is an arc. The proof of the next result is a composite of arguments in [**39**] and [**54**].

PROPOSITION 3.10. *Let (X, d) be an $\mathbb{R}$-metric space. The following are equivalent*

(1) *(X, d) is a $\mathbb{R}$-tree.*
(2) *Given two points of X, there is a unique arc having them as endpoints, and it is a segment.*
(3) *(X, d) is geodesic, and X contains no subspace homeomorphic to the circle S^1.*

PROOF. (1) $\Longrightarrow$ (2). Let α be a homeomorphism from a compact interval $[a, b]_{\mathbb{R}}$ into X, with image A, and let $x = \alpha(a)$, $y = \alpha(b)$. By Lemma 3.7, $[x, y] \subseteq A$. Hence $\alpha^{-1}\big([x, y]\big)$ is a connected subset of $[a, b]$ containing a and b, so is equal to $[a, b]$. It follows that $A = [x, y]$.

(2) $\Longrightarrow$ (3) If X contains a homeomorphic image of a circle, then the images of two semicircles give two distinct arcs joining the same two points of X.

(3) $\Longrightarrow$ (1) We have to verify Axioms (b) and (c) for an $\mathbb{R}$-tree. First, we show that if x, $y \in X$, there is a unique segment in X whose set of endpoints is $\{x, y\}$. For otherwise, using appropriate reflections and translations, there are isometries α, β: $[0, a] \to X$ with $\alpha(0) = \beta(0) = x$, $\alpha(a) = \beta(a) = y$, say, and $\alpha(b) \neq \beta(b)$ for some $b \subset [0, a]$. Let l be the largest element of $[0, b]$ on which α, β agree, and let m be the smallest element of $[b, a]$ on which α, β agree. Then $\alpha\big([l, m]\big)$, $\beta\big([l, m]\big)$ are two arcs meeting only at their endpoints, so by the "gluing lemma" of elementary topology we can define a continuous bijection from their union onto S^1; since their union is compact and S^1 is Hausdorff, this map is a homeomorphism, contrary to assumption.

Next, we show that Axiom (c) is satisfied. Let σ and τ be segments with a common endpoint v. Again we can find surjective isometries α: $[0, a]_{\mathbb{R}} \to \sigma$, β: $[0, b]_{\mathbb{R}} \to \tau$ with $\alpha(0) = \beta(0) = v$. Define $I = \big\{c \in \big[0, \min\{a, b\}\big] \mid \alpha(c) = \beta(c)\big\}$. Then $\alpha(I) = \beta(I) = \sigma \cap \tau$. For if $u \in \sigma \cap \tau$, then $u = \alpha(t) = \beta(t)$, where $t = d(v, u)$, and the reverse inclusion is obvious. Also, since $\mathbb{R}$ is Hausdorff, I is closed, so compact. If x, $y \in I$ and $z \in \Lambda$ with $x \leq z \leq y$, then $z \in I$, for otherwise the images of $\alpha|_{[x,y]_\Lambda}$ and $\beta|_{[x,y]_\Lambda}$ are distinct segments in X with the same endpoints, contrary to what has been proved. Thus I is a connected subspace of $\mathbb{R}$. Hence I is $[0, c]_{\mathbb{R}}$ for some c, and it follows that $\sigma \cap \tau$ is a segment.

It remains to verify Axiom (b). Suppose $[x, y] \cap [x, z] = \{x\}$. If $x \in [y, z]$, then it is easy to see that $[x, y] \cup [x, z] = [y, z]$. Suppose $x \notin [y, z]$, so

$$[y, z] \cap [y, x] = [y, w] \quad \text{with } w \neq x$$
$$[y, z] \cap [z, x] = [z, v] \quad \text{with } v \neq x.$$

It follows that each pair of the segments $[x, w]$, $[w, v]$ and $[v, x]$ meet only in a single endpoint. For example, since $w \in [x, y]$ and w, $v \in [y, z]$,

$$\begin{aligned}[x,w] \cap [w,v] &= [x,w] \cap [x,y] \cap [w,v] \\ &\subseteq [x,w] \cap [x,y] \cap [y,z] \\ &= [x,w] \cap [y,w] = \{w\}.\end{aligned}$$

It now follows as before that the union of the three segments is homeomorphic to S^1, a contradiction. This completes the proof.

Condition (3) of the last result may be viewed as an analogue of the definition of a simplicial tree, where circuits are replaced by homeomorphic images of a circle. In Proposition 3.3, we characterised Λ-trees using the notion of 0-hyperbolic space. For $\mathbb{R}$-trees, we have the following formulation.

PROPOSITION 3.11. *An $\mathbb{R}$-metric space (X, d) is an $\mathbb{R}$-tree if and only if*
(1) *it is connected,*
(2) *it is 0-hyperbolic.*

PROOF. An $\mathbb{R}$-tree is geodesic, so path connected and hence connected, and is 0-hyperbolic by 3.3. Conversely, assume (1) and (2). By Theorem 3.4 there is an embedding of (X, d) in an $\mathbb{R}$-tree (X', d'). Let x, $y \in X$, suppose $v \in X' \backslash X$ and $v \in [x, y]$. Then $[v, x] \not\equiv [v, y]$, and x, y are in different path components of $X' \backslash \{v\}$, by Lemma 3.8. Let C be the path component of $X' \backslash \{v\}$ containing x. We show that C is open and closed in $X' \backslash \{v\}$. For if u, $w \in X' \backslash \{v\}$, and $d(u, w) < d(u, v)$, then $d(u, z) < d(u, v)$ for all $z \in [u, w]$, so $v \notin [u, w]$. Hence the open ball B, centre u and radius $d(u, v)$ in $X' \backslash \{v\}$ is path connected. Thus if $u \in C$, $B \subseteq C$, and if $u \notin C$, then $B \cap C = \emptyset$. It follows that C and its complement are open in $X' \backslash \{v\}$, as required.

Since $X \subseteq X' \backslash \{v\}$, it follows that $X \cap C$ is open and closed in X. Since $x \in X \cap C$, $y \notin X \cap C$, this contradicts the connectedness of X. Thus $[x, y] \subseteq X$, and (X, d) is geodesic. It follows that (X, d) is an $\mathbb{R}$-tree by Proposition 3.3.

4. Isometries of Λ-Trees

We begin by considering the behaviour of a single isometry of a Λ-tree. There is a classification of such isometries similar to that for hyperbolic space (in fact there is such a classification for any proper geodesic hyperbolic $\mathbb{R}$-metric space in the sense of Gromov-see [**19**] or [**29**]). There are three possible kinds of isometries in a Λ-tree, elliptic, hyperbolic and inversions. Inversions do not occur in the case $\Lambda = \mathbb{R}$, or indeed whenever $\Lambda = 2\Lambda$, and unlike hyperbolic space, there are no parabolic isometries.

After this discussion we shall look briefly at a classification of group actions on Λ-trees. We shall also describe an interesting method, due to Serre [**47**; Chapter II §1] in the case $\Lambda = \mathbb{Z}$ and to Morgan and Shalen [**39**] in general, of constructing an action of $\mathrm{SL}_2(F)$ on a Λ-tree, where F is a field with a valuation. The rest of the article is a survey of results on free actions on Λ-trees.

Let (X, d) be a Λ-tree, where Λ is an arbitrary ordered abelian group, and let g be an isometry of X onto X. We call g *elliptic* if it has a fixed point. The behaviour of such isometries is described in the next lemma, where $\langle g \rangle$ denotes the cyclic subgroup generated by g in the group of metric automorphisms of X. If $[x, y]$ is a segment, there is a unique point p such that $d(x,p) = d(p,y)$ if and only if $d(x,y) \in 2\Lambda$. This point, when it exists, is called the midpoint of $[x,y]$.

LEMMA 4.1. *Let g be an elliptic isometry and let X^g denote the set of fixed points of g. Then*

(1) *X^g is a closed non-empty $\langle g\rangle$-invariant subtree of X which meets every $\langle g\rangle$-invariant subtree of X.*
(2) *X^g is contained in every subtree of X with the property that it meets every $\langle g\rangle$-invariant subtree of X.*
(3) *If $x \in X$ and $[x,p]$ is the bridge between x and X^g, then p is the midpoint of $[x, gx]$, and for all $a \in X^g$, $[a,x] \cap [a, gx] = [a,p]$. Hence $d(x, gx) = 2d(x,p)$.*

PROOF. See [**1**; Proposition 6.1].

LEMMA 4.2. *The following are equivalent*

(1) *There is a segment of X invariant under g, and the restriction of g to this segment has no fixed points.*
(2) *There is a segment $[x,y]$ in X such that $gx = y$, $gy = x$ and $d(x,y) \notin 2\Lambda$.*
(3) *g^2 has a fixed point, but g does not.*
(4) *g^2 has a fixed point, and for all $x \in X$, $d(x, gx) \notin 2\Lambda$.*

PROOF. For the implications (1) $\Longrightarrow$ (2) and (2) $\Longrightarrow$ (3) see [**1**], Proposition 6.3. For (3) $\Longrightarrow$ (4) see [**14**], Lemma 4.2.

(4) $\Longrightarrow$ (1): Let p be a fixed point of g^2. Then $[p, gp]$ is invariant under g, and g has no fixed points at all, by Lemma 4.1(3).

DEFINITION. If g satisfies the equivalent conditions of 4.2, it is called an inversion.

Examples of inversions are provided by certain reflections, e.g. take $X = \Lambda = \mathbb{Z}$ and $g\colon x \mapsto 1 - x$. In fact, inversions in $\mathbb{Z}$-trees correspond to graph automorphisms of the corresponding simplicial tree which interchange e and $\bar{e}$ for some edge e, so the terminology is consistent with that of §2.

DEFINITION. An isometry that is not an inversion and not elliptic is called a hyperbolic isometry.

A subtree of a Λ-tree is called *linear* if it is metrically isomorphic to a subtree of Λ.

We define

$$A_g = \{p \in X \mid [g^{-1}p, p] \cap [p, gp] = \{p\}\}$$

and call A_g the *characteristic subset* of g. If g is elliptic, it follows from 4.1 that $A_g = X^g$. If g is an inversion, then $A_g = \emptyset$. For suppose $p \in A_g$, g has no fixed point but g^2 does. It follows from 4.1 that the midpoint p of $[g^{-1}p, gp]$ is fixed by g^2. But then $g^{-1}p = gp$, and since $p \in [g^{-1}p, gp]$, $p = gp$, a contradiction.

THEOREM 4.3. *Suppose g is hyperbolic. Then*

(1) *A_g is a non-empty closed $\langle g\rangle$-invariant subtree of X which meets every $\langle g\rangle$-invariant subtree of X.*
(2) *A_g is contained in every subtree of X with the property that it meets every $\langle g\rangle$-invariant subtree of X.*

(3) *A_g is a linear tree, and g restricted to A_g is equivalent to a translation $a \mapsto a + \ell(g)$ for some $\ell(g) \in \Lambda$ with $\ell(g) > 0$.*

(4) *If $x \in X$ and $[x,p]$ is the bridge between x and A_g, then $[x, gx] \cap A_g = [p, gp]$ and $d(x, gx) = \ell(g) + 2d(x,p)$.*

PROOF. See [**1**] Theorem 6.6.

When g is hyperbolic A_g is called the axis of g (which explains the notation). When g is elliptic or an inversion, we put $\ell(g) = 0$. Thus $\ell(g)$ is defined for any g, and it is called the hyperbolic length of g.

COROLLARY 4.4. *Assume g is not an inversion. Then*

$$\ell(g) = \min\{d(x, gx) \mid x \in X\} = \min_{x \in X} L_x(g)$$

and $A_g = \{p \in X \mid d(p, gp) = \ell(g)\}$. Further, for any $x \in X$,

$$\ell(g) = \min\{L_x(g^2) - L_x(g), 0\}.$$

PROOF. The first two parts are clear if g is elliptic and follow from 4.3(iv) if g is hyperbolic.

Suppose g is elliptic. From Lemma 4.1 we have $L_x(g) = 2d(x,p)$ where p is a point stabilized by g, so by g^2. By the triangle inequality, $L_x(g^2) \leq d(x,p) + d(p, g^2x) = d(x,p) + d(g^2p, g^2x) = 2d(x,p)$, so $\max\{L_x(g^2) - L_x(g), 0\} = 0$, as required.

Finally, if g is hyperbolic, let $[p, x]$ be the bridge between x and A_g. By Theorem 4.3, $L_x(g) = \ell(g) + 2d(p, gp)$. It is a consequence of Theorem 4.3 that $A_{g^2} = A_g$, $\ell(g^2) = 2\ell(g)$ and g^2 is hyperbolic (see 6.13 in [**1**]). Hence $L_x(g^2) = 2\ell(g) + 2d(p, gp)$, so $L_x(g^2) - L_x(g) = \ell(g) > 0$ and the result follows.

REMARK. The final formula in Corollary 4.4 is also true when g is an inversion; see 6.13 in [**1**] (which uses the "base change" functor), or Lemma 4.5 in [**14**]. If g, h are isometries of a Λ-tree onto itself, then g is an inversion if and only if $A_g = \emptyset$, and g is an inversion if and only if hgh^{-1} is an inversion. It therefore follows from Corollary 4.4 that for all isometries g, h, $A_{hgh^{-1}} = hA_g$ and $\ell(hgh^{-1}) = \ell(g)$.

Classification of actions. If G is a group acting on a Λ-tree (X, d), there is a mapping $\ell: G \to \Lambda$, where $\ell(g)$ is the hyperbolic length of the isometry of X induced by g. We call ℓ the hyperbolic length function for the action. Actions on Λ-trees can be classified into three types; there are various ways of characterizing the different types, and we shall give just one of them. An action of G on a Λ-tree is

(1) *abelian* if $\ell(gh) \leq \ell(g) + \ell(h)$ for all g, $h \in G$;

(2) *dihedral* if it is non-abelian, but for all g, $h \in G$ with $\ell(g) > 0$, $\ell(h) > 0$, we have $\ell(gh) \leq \ell(g) + \ell(h)$;

(3) *irreducible* if $\ell(gh) > \ell(g) + \ell(h)$ for some g, $h \in G$ with $\ell(g) > 0$, $\ell(h) > 0$.

A geometric interpretation of the defining conditions is given by Theorem 4.5 below. If g, h are non-inversions, then $\ell(gh) \leq \ell(g) + \ell(h)$ if and only if A_g and A_h intersect.

An example of an abelian action is obtained by taking $X = \Lambda$, with Λ acting on itself by translations ($ax = x + a$ for a, $x \in \Lambda$). The full group of metric automorphisms of Λ consists of the translations and reflections, and is a split extension of the group of translations (which is isomorphic to Λ) by a cyclic group of order 2 generated by a single reflection. See 2.1 in [**1**]. It is easy to see that this action is dihedral. In the case $\Lambda = \mathbb{Z}$, the group is the infinite dihedral group, which partially explains the terminology.

An example of an irreducible action is obtained by taking the action of the free group of rank two on its Cayley graph, relative to a basis which we denote by $\{a, b\}$ (see the discussion after Corollary 2.3). The elements a and bab^{-1} have axes $A_a = \{\dots a^{-2}, a^{-1}, 1, a, a^2 \dots\}$ and $A_{bab^{-1}} = bA_a$, which do not intersect. It follows that the action is irreducible by Theorem 4.5 below.

We shall not say much more about the classification, except to note that abelian actions can be further classified into different types. We refer to [**1**; §7] for further information. However, it is worth noting that there is a theory of ends of Λ-trees, analogous to that for locally compact spaces. We shall not say anything about this important aspect of the theory of Λ-trees. There is an account in [**1**], and the theory is further developed in [**3**]. Any action of a group by isometries induces an action on the ends, and any action with a fixed end is necessarily abelian. An action with a fixed end such that $\bigcap_{g\in G} A_g = \emptyset$ is called an action of end type. For $\Lambda = \mathbb{Z}$, these are related to decompositions of a group as ascending HNN-extensions, via the Bass-Serre theory. For $\Lambda = \mathbb{R}$, they are related, in the case of a finitely generated group, to the Bieri-Neumann-Strebel invariant of the group. These matters are discussed in [**10**].

The terminology "irreducible" is taken from [**20**]; in [**1**], such actions are called actions of general type. By combining arguments in [**1**] and [**20**], one can show that an action of G is abelian if and only if there is a homomorphism $\tau: G \to \Lambda$ such that $\ell(g) = |\tau(g)|$ for all $g \in G$, and this is equivalent to: $\ell([g, h]) = 0$ for all g, $h \in G$, where $[g, h]$ denotes the commutator of g and h. (See [**15**], Proposition 7).

THEOREM 4.5. *Suppose G acts as isometries on a Λ-tree (X, d) and g, $h \in G$ do not act as inversions. Then*

$$d(A_g, A_h) = \max\left\{\frac{1}{2}(\ell(gh) - \ell(g) - \ell(h)), 0\right\}.$$

PROOF. Suppose $A_g \cap A_h = \emptyset$. Let $[p, q]$ be the bridge between A_g and A_h, with $p \in A_g$, $q \in A_h$. Then it is shown in §8 of [**1**] that $[p, q] \subseteq A_{gh}$, so gh is not an inversion, and that

$$\ell(gh) = \ell(g) + \ell(h) + 2d(A_g, A_h).$$

We shall not give the details. However, it is helpful to see the pictures illustrating the proof, and in Fig. 4 we reproduce one of them, illustrating the case $\ell(g) > 0$, $\ell(h) > 0$. The double line indicates the axis of gh. Note that $p \in A_{gh}$, so $\ell(gh) = d(p, ghp)$, which can be read off from the picture. This is justified using 2.14 in [**1**].

Now suppose $A_g \cap A_h \neq \emptyset$. Take $p \in A_g \cap A_h$. Then $\ell(gh) \leq d(p, ghp)$, either by definition if gh is an inversion, or by Corollary 4.4 otherwise. By the triangle inequality,

$$d(p, ghp) \leq d(p, gp) + d(gp, ghp) = d(p, gp) + d(p, hp) = \ell(g) + \ell(h)$$

again by Corollary 4.4. The theorem follows.

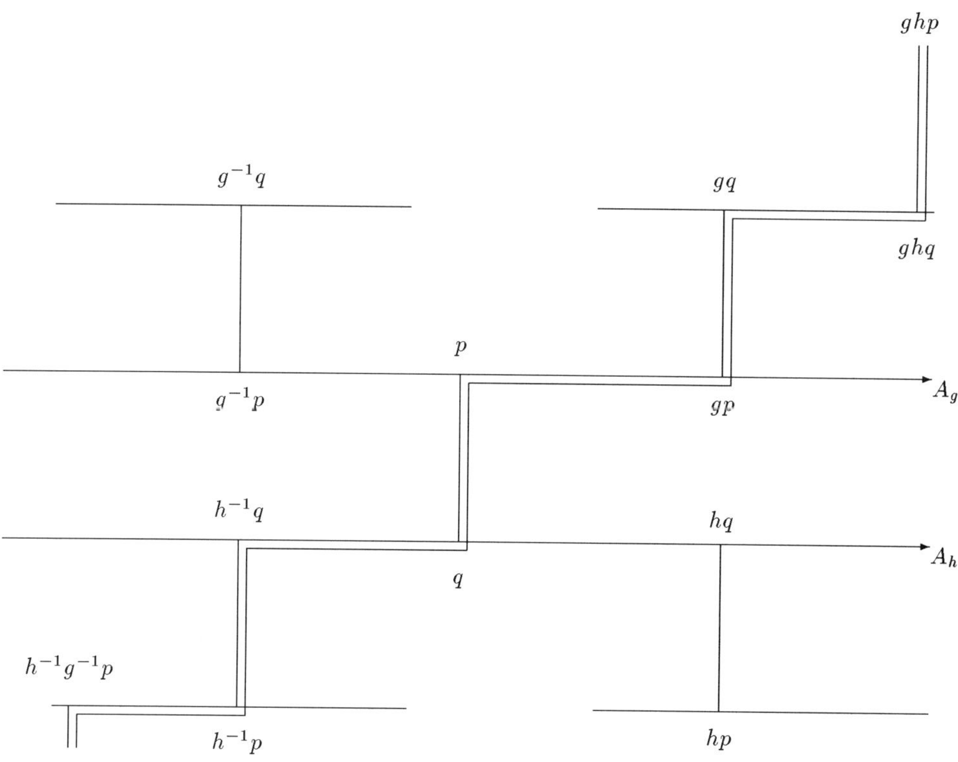

FIGURE 4. The axis of gh in one of the cases of 4.5.

The action of $\mathrm{GL}_2(F)$. Let F be a field and let $F^\times = F\backslash\{0\}$ be the multiplicative group of F. The idea of a valuation on F is standard.

DEFINITION. A valuation on F is a group homomorphism $v\colon F^\times \to \Gamma$, where Γ is an ordered abelian group, such that, for all x, $y \in F$, $v(x+y) \geq \min\{v(x), v(y)\}$, where we put $v(0) = \infty$ with the convention that $\lambda < \infty$ for all $\lambda \in \Lambda$.

Given such a valuation, let $\Lambda = v(F^\times)$, the image of v. We call Λ the value group of the valuation. The set $A = \{a \in F \mid v(a) \geq 0\}$ is a subring of F, called the valuation ring of v. It is a local ring with maximal ideal $\{a \in F \mid v(a) > 0\}$.

Let V be the vector space F^2. An A-lattice in V is a finitely generated A-module $L \subseteq V$ which spans V as an F-module. Given lattices L, L', we define $L \sim L'$ to mean that $L' = aL$ for some $a \in F$. It can be shown that this is an equivalence relation on the set of A-lattices in V and we denote the set of equivalence classes by X_v. The equivalence class of L is denoted by $\bar{L}$. Further, one can show that two equivalence classes can be expressed as $\bar{L}_0$, $\bar{L}_1$, where $L_0 \subseteq L_1$ and L_1/L_0 is isomorphic to a cyclic A-module A/cA for some $c \in A$. We put $d(\bar{L}_0, \bar{L}_1) = v(c)$, and this gives a well-defined function from $X_v \times X_v$ to Λ. For proofs of all these assertions and the next theorem, see Appendix A in [**1**].

THEOREM 4.6. *The pair* (X_v, d) *is a* Λ*-tree.*

The natural action of $\mathrm{GL}(V) = \mathrm{GL}_2(F)$ on V induces an action on X_v as isometries. Put $L_0 = A^2$, and if $g = \begin{pmatrix} a & b \\ c & d \end{pmatrix} \in \mathrm{GL}_{2(F)}$, define

$$v(g) = \min\{v(a), v(b), v(c), v(d)\}.$$

The next result gives the Lyndon length function based at $\bar{L}_0$ and the hyperbolic length function ℓ for this action.

Lemma 4.7. *For $g \in \mathrm{GL}(V)$,*
(1) $d(g\bar{L}_0, \bar{L}_0) = v(\det(g)) - 2v(g)$;
(2) $\ell(g) = \max\{v(\det(g)) - 2v(\mathrm{tr}(g)), 0\}$.

Proof. See A.10 and B.7 in [**1**].

For $g \in \mathrm{SL}_2(F)$, the lemma gives $d(g\bar{L}_0, \bar{L}_0) = -2v(g) \in 2\Lambda$, so by criterion (4) of Lemma 4.2 the action of $\mathrm{SL}_2(F)$ on X_v is without inversions. Hence $g \in \mathrm{SL}_2(F)$ has a fixed point if and only if $\ell(g) = 0$, which by Lemma 4.7 happens if and only if $v(\mathrm{tr}(g)) \geq 0$. Thus g acts as a hyperbolic isometry if and only if $v(\mathrm{tr}(g)) < 0$.

Free actions. We call a group Λ-free if it has a free action without inversions on some Λ-tree. Free means that the stabilizer of every point is trivial. Thus an action of a group G on a Λ-tree is free and without inversions if and only if $\ell(g) > 0$ for all $g \in G$ with $g \neq 1$. We call a group tree-free if it is Λ-free for some ordered abelian group Λ. We give some examples.

Example 1. A group is $\mathbb{Z}$-free if and only if it is a free group, by Corollary 2.3. This is the simplest case of the Bass-Serre theory, and it is this that suggests the study of free actions on Λ-trees may be simpler than the study of actions in general.

Example 2. Subgroups of Λ are Λ-free, since they act on Λ as translations. In particular, finitely generated free abelian groups are $\mathbb{R}$-free.

Example 3. Let G be the fundamental group of a closed surface. If the surface is orientable, this means G has a presentation of the form

$$G = \langle a_1, b_1, \ldots, a_g, b_g \mid [a_1, b_1] \ldots [a_g, b_g] = 1 \rangle$$

and otherwise G has a presentation of the form

$$G = \langle x_1, \ldots, x_g \mid x_1^2 \ldots x_g^2 = 1 \rangle.$$

In the first case we call G orientable, in the second case G is called non-orientable, and in both cases g is called the genus. It has been shown by Morgan and Shalen [**40**] that if G is orientable, or non-orientable of genus greater than 3, then G is $\mathbb{R}$-free. This is a difficult result, proved using the theory of measured geodesic laminations. Further, Morgan and Shalen show that the non-orientable surface groups of genus at most 3 are not $\mathbb{R}$-free. (This was also noted by W. Parry). We shall call these three groups *exceptional surface groups*.

Example 4. If G_i are Λ-free groups for all $i \in I$, then the free product $\ast_{i \in I} G_i$ is Λ-free. This was essentially proved by Harrison [**31**] using Lyndon length functions.

EXAMPLE 5. In [**3**], Bass has examined the structure of groups acting on Λ-trees, where Λ has the form $\mathbb{Z} \oplus \Lambda_0$ with the lexicographic ordering, Λ_0 being an arbitrary ordered abelian group. The results are complicated to state, and we shall confine attention to some simple special cases of tree-free groups arising from this work. If A and B are free groups, $a \in A$, $b \in B$, a is not a proper power in A, b is not a proper power in B, and $a \neq 1 \neq b$, then the free product with amalgamation

$$A \underset{\langle a \rangle = \langle b \rangle}{*} B$$

is $(\mathbb{Z}\oplus\mathbb{Z})$-free, where $\mathbb{Z}\oplus\mathbb{Z}$ has the lexicographic ordering. Also, if A is a free group, a, $b \in A$, a and b are not proper powers in A, $a \neq 1 \neq b$ and a is not conjugate to b^{-1} in A, then the HNN-extension

$$\langle t, A \mid \mathrm{rel}(A), tat^{-1} = b \rangle$$

is $(\mathbb{Z} \oplus Z)$-free.

Our final example is that of non-standard free groups, and this merits a section on its own. Before that, we shall consider a theorem of Rips which characterizes finitely generated $\mathbb{R}$-free groups. First, we shall prove certain necessary results on residual properties of some of the examples above.

DEFINITION. A group G is called n-residually free, where n is a positive integer, if given a collection $g_1, \ldots, g_n$ of elements of $G\backslash\{1\}$, there is a homomorphism $\phi: G \to F$, where F is a free group, such that $\phi(g_i) \neq 1$ for $1 \leq i \leq n$. A group is called fully residually free (or ω-residually free) if it is n-residually free for all $n > 0$.

Thus "1-residually free" just means residually free, with the usual terminology.

LEMMA 4.8. *Finitely generated free abelian groups, and the non-exceptional surface groups are fully residually free.*

PROOF. First, we show that if A, B are abelian fully residually free groups, then $A \times B$ is fully residually free. Let $(a_1, b_1), \ldots, (a_n, b_n)$ be non-identity elements of $A \times B$. Since A, B are abelian, there are homomorphisms $\phi: A \to \mathbb{Z}$, $\psi: B \to \mathbb{Z}$ (viewing $\mathbb{Z}$ as an infinite cyclic group) such that

$$\phi(a_i) \neq 0 \quad \text{for all } i \text{ such that } a_i \neq 1$$
$$\psi(b_i) \neq 0 \quad \text{for all } i \text{ such that } b_i \neq 1.$$

Then $\theta = \phi \times \psi: A \times B \to \mathbb{Z} \times \mathbb{Z}$ is a homomorphism with $\theta(a_i, b_i) \neq 0$ for $1 \leq i \leq n$. It therefore suffices to show that $\mathbb{Z} \times \mathbb{Z}$ is fully residually free. Let $(c_1, d_1), \ldots, (c_n, d_n)$ be non-identity elements of $\mathbb{Z} \times \mathbb{Z}$. Let x be a prime not occurring in the factorization of any non-zero c_i or d_i. The homomorphism $f: \mathbb{Z} \times \mathbb{Z} \to \mathbb{Z}$ given by $f(1,0) = x$, $f(0,1) = 1$ then satisfies $f(c_i, d_i) \neq 0$ for $1 \leq i \leq n$, as required. It follows by induction on m that the free abelian group $\mathbb{Z}^m$ is fully residually free.

The surface group with presentation $\langle x_1, \ldots, x_g \mid x_1^2, \ldots, x_g^2 = 1 \rangle$ $(g \geq 4)$ is fully residually free by §7 in [**4**]. This group has the surface group $\langle a_1, b_1, \ldots, a_{g-1}, b_{g-1} \mid [a_1, b_1] \ldots [a_{g-1}, b_{g-1}] = 1 \rangle$ as a subgroup of index 2 (see §2.2 in [**59**]). Since the class of fully residually free groups is clearly closed under taking subgroups, the group $\langle a_1, b_1, \ldots, a_g, b_g \mid [a_1, b_1], \ldots, [a_g, b_g] = 1 \rangle$ $(g \geq 3)$ is fully residually free.

For $g = 2$, the group $\langle a_1, b_1, a_2, b_2 \mid [a_1, b_1][a_2, b_2] = 1\rangle$ is a free product with amalgamation $F \underset{u=\bar{u}}{*} \overline{F}$, where F is the free group with basis $\{a_1, b_1\}$, $\overline{F}$ the free group on $\{a_2, b_2\}$, isomorphic to F via the isomorphism sending a_1 to b_2 and b_1 to a_2, $u = [a_1, b_1]$ and $\bar{u}$ is the image of u under this isomorphism. Further u is not a proper power in F. It follows from Theorem 8 in [**4**], using the observations in §1.3 of [**5**], that this surface group is fully residually free. Finally, for $g = 1$, $\langle a_1, b_1 \mid [a_1, b_1] = 1\rangle$ is actually free abelian of rank 2.

It is easy to deduce that any direct product of infinite cyclic groups is fully residually free, as observed in §4 of [**4**].

COROLLARY 4.9. *Free abelian groups and surface groups are residually finite.*

PROOF. Free abelian groups and the orientable surface groups are residually free by 4.8 and the remark following the proof, so are residually finite since free groups are residually finite (see Proposition 5, Chapter 1 in [**17**]). By §2.2 in [**59**], the non-orientable surface groups of genus at least 2 contain an orientable surface group as a subgroup of index 2, so are residually finite. (We are using the elementary fact that the class of residually finite groups is closed under finite extensions; see 26.12 in [**42**]). Finally the non-orientable surface group of genus 1 is cyclic of order 2, so is finite.

One final remark on surface groups is needed.

REMARK. Abelianising $\langle a_1, b_1, \ldots, a_g, b_g \mid [a_1, b_1] \ldots [a_g, b_g] = 1\rangle$ gives a free abelian group of rank $2g$, while abelianising $\langle x_1, \ldots, x_g \mid x_1^2 \ldots x_g^2 = 1\rangle$ gives the direct product of a free abelian group of rank $g - 1$ and a cyclic group of order 2. It follows that the minimum number of generators of $\langle a_1, b_1, \ldots, a_g, b_g \mid [a_1, b_1], \ldots, [a_g, b_g] = 1\rangle$ is $2g$, and the minimum number of generators of

$$\langle x_1, \ldots, x_g \mid x_1^2, \ldots, x_g^2 = 1\rangle$$

is g.

Rips' theorem. According to the preceding examples, a free product of groups each of which is either finitely generated free abelian or a non-exceptional surface group, is $\mathbb{R}$-free. Rips' Theorem gives a converse for finitely generated groups.

RIPS' THEOREM. *A finitely generated $\mathbb{R}$-free group is a free product of finitely many finitely generated free abelian groups and surface groups.*

Rips outlined a proof of this at a conference at the Isle of Thorns, a conference centre of the University of Sussex, England, in 1991. It was not published, but versions of it have now appeared in [**7**] and [**25**]. We cannot give the proof here, but will indicate some ideas in the proof in [**25**]. The main idea had been considered earlier by Makanin and Razborov, and is contained in the next definition.

DEFINITION. A *system of isometries* $Z = (D, \Phi)$ consists of a union D of finitely many disjoint compact intervals in $\mathbb{R}$, with a finite collection $\Phi = \{\phi_j \mid 1 \leq j \leq q\}$, where $\phi_j \colon A_j \to B_j$ is an isometry and A_j, B_j are closed subintervals of the intervals in D.

Note that the intervals A_j, B_j may consist of a single point. We can associate to such a system Z a topological space Σ as follows. Take the disjoint union of D

and $\bigcup_{j=1}^{k}(A_j \times [0,1])$, and let Σ be the quotient space obtained by identifying $(t,0)$ in $A_j \times [0,1]$ with $t \in A_j \subseteq D$, and $(t,1)$ in $A_j \times [0,1]$ with $\phi_j(t) \in B_j \subseteq D$, for every $t \in A_j$ and $1 \leq j \leq q$.

We let $\sim$ be the equivalence relation on Σ generated by the relation $\simeq$, where $x \simeq y$ if and only if there exist j and $a \in A_j$ such that x, y belong to the image of $\{a\} \times [0,1]$ in Σ. The set of equivalence classes is called the foliation of Σ and is denoted by $\mathcal{F}$. The equivalence classes are called the *leaves* of the foliation.

The system Z is called connected if Σ is connected. If Z is connected, we can associate a certain quotient of $\pi_1(\Sigma)$ with Z, denoted by $G(Z)$, and the main part of the proof consists of showing that $G(Z)$ has the form claimed in Rips' Theorem, i.e. a free product of free abelian groups and surface groups. This, is however, long and difficult, and we shall not attempt to describe it. (For those who understand the necessary topology, $G(Z)$ is the quotient of $\pi_1(\Sigma)$ by the normal subgroup generated by the free homotopy classes of loops contained in the leaves of Σ. Also, Σ is a 2 dimensional CW complex, with the homotopy type of a finite topological graph, i.e. a 1-dimensional CW-complex).

Now suppose G is a group acting on an $\mathbb{R}$-tree (X,d), and G is generated by the finite set $\{s_1,\ldots,s_m\}$. Let K be a subtree of X spanned by a finite set of points. Let D be the disjoint union $I_1\cup,\ldots,\cup I_n$, where $I_1,\ldots,I_n$ are the closures of the components of $K\backslash B$, and B is the set of branch points of K (i.e. points of degree greater than 2 in K). We view D as embedded in $\mathbb{R}$ so that the intervals $I_1,\ldots,I_n$ are pairwise disjoint, but the embedding can be arbitrary. If $b \in B$, and $I_{i_1},\ldots,I_{i_l}$ are the segments having b as an endpoint, let $x_{i_1}^b,\ldots,x_{i_l}^b$ be the corresponding endpoints of the segments $I_{i_1},\ldots,I_{i_l}$ in D. Define $\Phi_0 = \bigcup_{b\in B}\{(x_{i_1}^b \mapsto x_{i_2}^b),\ldots,(x_{i_1}^b \mapsto x_{i_l}^b)\}$, where $(x \mapsto y)$ indicates the isometry with domain $\{x\}$, mapping x to y.

Each generator s_i defines an isometry $\theta_i\colon K\cap s_i^{-1}(K) \to s_iK\cap K$ by restriction. Further, each θ_i defines by restriction an isometry ϕ_{ijk} from a closed subinterval of I_j to a closed subinterval of I_k. (Restrict θ_i to $I_i \cap s_i^{-1}(I_k)$). Let $\Phi = \Phi_0 \cup \{\phi_{ijk} \mid 1 \leq i \leq m,\ 1 \leq j,k \leq n\}$, to obtain a system of isometries (D,Φ) which we denote by $Z(K)$. (Any ϕ_{ijk} with empty domain can be ignored). Because $\Phi_0 \subseteq \Phi$, $Z(K)$ is connected. We write $G(K)$ for $G\big(Z(K)\big)$.

Now G is countable, say $G = \{g_i \mid i \geq 1\}$, and if $v \in X$, then it follows from Lemma 3.6 that $\bigcup_{i=1}^{\infty}[v,g_iv]$ is a G-invariant subtree of X, so we can assume that $X = \bigcup_{i=1}^{\infty}[v,g_iv]$. Also, let $K_r = \bigcup_{i=1}^{r}[v,g_iv]$, the subtree spanned by v and $g_1v,\ldots,g_rv$, so $X = \bigcup_{r=1}^{\infty}K_r$. For sufficiently large r, it is not difficult to show that $G(K_r)$ has a presentation with generators $g_1,\ldots,g_m$ and relations the words $g_{i_1}^{e_1},\ldots,g_{i_p}^{e_p}$ (where $e_i = \pm1$) such that there exists $x \in K_r$ with $g_{i_1}^{e_1},\ldots,g_{i_p}^{e_p}(x) = x$ and $g_{i_j}^{e_j},\ldots,g_{i_p}^{e_p}(x) \in K_r$ for $1 \leq j \leq p$. It follows that for sufficiently large r, there are epimorphisms $\rho_r\colon G(K_r) \to G(K_{r+1})$ and epimorphisms $\mu_r\colon G \to G(K_r)$ with $\mu_{r+1} = \rho_r\mu_r$, so there is an epimorphism of G onto the direct limit of the sequence (ρ_r). The kernel is the (normal) subgroup generated by the set of elliptic elements of G, so if G acts freely then G is isomorphic to the direct limit.

Every $G(K_r)$ is a free product of surface groups and free abelian groups, so is residually finite by 4.9 and the fact that free products of residually finite groups are residually finite (see the corollary to Proposition 22, Chapter 1 in [**17**]). All the $G(K_r)$ are therefore Hopfian, by a well-known result of Mal'cev (see the corollary to Proposition 5, Chapter 1 in [**17**]). Further, since $G(K_r)$ is generated by m elements for sufficiently large r, it follows from the remark after Corollary 4.9 and

Grushko's Theorem that only finitely many isomorphism classes of groups appear in the sequence $\big(G(K_r)\big)_{r\geq 1}$. It follows that ρ_r is an isomorphism for sufficiently large r, so the direct limit is isomorphic to $G(K_r)$ for sufficiently large r. Rips' Theorem follows.

It is not true that if G is an $\mathbb{R}$-free group which is not finitely generated, then G is a free product of free abelian groups and surface groups. See [**23**] for a counterexample.

Non-standard free groups. Our final example of a class of tree-free groups is of non-standard models of the free groups, and their subgroups. We shall only consider such models arising as ultrapowers of free groups, and begin by describing this construction. Let I be a set and let 2^I be the Boolean algebra of all subsets of I. An ultrafilter on I is a subset $\mathcal{D}$ of 2^I such that

(i) $A \in \mathcal{D}$ and $A \subseteq B \subseteq I$ implies $B \in \mathcal{D}$.

(ii) $A, B \in \mathcal{D}$ implies $A \cap B \in \mathcal{D}$.

(iii) For all $A \in 2^I$, exactly one of A, $I \backslash A$ belongs to $\mathcal{D}$.

Let $\{X_i \mid i \in I\}$ be a family of sets indexed by I, and let $\mathcal{D}$ be an ultrafilter on I. The ultraproduct $\prod_{i\in I} X_i/\mathcal{D}$ is defined to be the quotient set $\prod_{i\in I} X_i/\!\sim$, where $(x_i)_{i\in I} \sim (y_i)_{i\in I}$ if and only if $\{i \in I \mid x_i = y_i\} \in \mathcal{D}$. This is easily checked to be an equivalence relation. We shall denote the equivalence class of an element $(x_i)_{i\in I} \in \prod_{i\in I} X_i$ by $\langle x_i\rangle_{i\in I}$, and usually abbreviate this to $\langle x_i\rangle$. If all $X_i = X$, a single set, then the ultraproduct is $X^I/\mathcal{D}$, which is called an ultrapower of X.

Note that X embeds in $X^I/\mathcal{D}$, by mapping x to $\langle x_i\rangle$, where $x_i = x$ for all $i \in I$. An ultrafilter $\mathcal{D}$ is called *principal* if it contains a finite set. We leave it as an exercise to show the embedding of X into $X^I/\mathcal{D}$ is a bijection if $\mathcal{D}$ is principal. Thus to obtain an ultrapower which could genuinely be described as a "non-standard extension" of X, we should assume $\mathcal{D}$ is non-principal. However, this is unnecessary in what follows and we shall not do so.

Denoting the ultrapower $X^I/\mathcal{D}$ by ${}^*X$, given a function $f: X^n \to Y$ of n variables, we obtain an extension to a function ${}^*f: ({}^*X)^n \to {}^*Y$ by defining ${}^*f(x^{(1)}, \ldots, x^{(n)}) = \big\langle f(x_i^{(1)}, \ldots, x_i^{(n)})\big\rangle$, where $x^{(1)} = \langle x_i^{(1)}\rangle$, etc. Similarly any finitary relations on X can be extended to ${}^*X$. Thus if X is a group, then we can define a multiplication in ${}^*X$ by $\langle x_i\rangle\langle y_i\rangle = \langle x_i y_i\rangle$, and this makes ${}^*X$ into a group. If X is abelian, so is ${}^*X$, and if X is an ordered abelian group, then ${}^*X$ can be made into an ordered abelian group by defining $\langle x_i\rangle \leq \langle y_i\rangle$ if and only if $\{i \in I \mid x_i \leq y_i\} \in \mathcal{D}$. Similarly if X is a ring or a field, so is ${}^*X$. These assertions can be proved directly, but all our claims follow from Łoš's Theorem (see Corollary 2.2 or Lemma 2.3 in [**6**], or for those not familiar with first order logic, [**56**]).

In Example 1 above, a free group F acts freely on the $\mathbb{Z}$-tree corresponding to its Cayley graph, relative to some basis X (see the remarks after Corollary 2.3). Using the vertex 1 as basepoint, according to the remarks preceding Theorem 3.5, the corresponding Lyndon length function L is given by: $L(u)$ is the length of the reduced word on $X^{\pm 1}$ representing u, and $c(u, v)$ is the length of the longest common initial segment of the reduced words representing u and v. Since the action is free and without inversions, $L(u^2) > L(u)$ for all $u \in F \backslash \{1\}$ by Corollary 4.4. Let I be an index set, let $\mathcal{D}$ be an ultrafilter in 2^I and denote the ultrapower $X^I/\mathcal{D}$ by ${}^*X$. Then as indicated above we obtain a function ${}^*L$: ${}^*F \to {}^*\mathbb{Z}$, given by ${}^*L\big(\langle u_i\rangle\big) = \big\langle L(u_i)\big\rangle$, and ${}^*L$ is a Lyndon length function. This can be checked directly, or one can appeal to a suitable version of Łoš's Theorem. Further,

${}^*L(u^2) > {}^*L(u)$ for all $u \in {}^*F\backslash\{1\}$. Therefore there is an action of ${}^*F$ on a ${}^*\mathbb{Z}$-tree by Theorem 3.5, and this action is free by Corollary 4.4 and the remark following it. Thus ${}^*F$, and all its subgroups, are tree-free.

This example is due to Gaglione and Spellman [**26**], [**27**] (see also Remeslennikov [**46**]). These authors were interested in groups with the same universal theory as the non-abelian free groups, and these are precisely the non-abelian subgroups of ultrapowers of F_2. This follows from Theorems 5.1 and 5.2 in [**14**]. Our next result will be a statement of the first of these theorems, giving a group-theoretic characterization of the subgroups of ultrapowers of F_2. The main point in the proof is that finitely generated subgroups of ultrapowers of F_2 are fully residually free, which is due to Remeslennikov [**45**]. We shall not give any details (a complete proof is given in [**14**]).

THEOREM 4.10. *Let G be any group. Then G embeds in some ultrapower ${}^*F_2$ of the free group of rank 2 if and only if G is locally fully residually free.*

We present another argument by Remeslennikov in [**46**], which is an interesting application of the action of $\mathrm{SL}_2(F)$ on a Λ-tree, where F is a field with a valuation, and Λ is the value group.

THEOREM 4.11. *If G is a finitely generated fully residually free group, then G is Λ-free for some finitely generated ordered abelian group Λ.*

PROOF. We shall not give all the details of the proof. Let $P = \mathbb{Q}(x_1, x_2, y_1, y_2)$ be a field extension of $\mathbb{Q}$ by four algebraically independent elements. There is a valuation v of P given by $v(f/g) = \deg(g) - \deg(f)$, where f, g are polynomials in x_1, x_2, y_1, y_2 and deg means total degree. Define

$$A = \begin{pmatrix} x_1 & x_1x_2 - 1 \\ 1 & x_2 \end{pmatrix} \qquad B = \begin{pmatrix} y_1 & y_1y_2 - 1 \\ 1 & y_2 \end{pmatrix}$$

so that $A, B \in \mathrm{SL}_2(P)$. Let H be the subgroup generated by A and B. If w is a reduced word on $\{A^{\pm 1}, B^{\pm 1}\}$, let $\|w\|$ denote the length of the cyclically reduced word corresponding to w. Remeslennikov shows that, if h is the element of H represented by w, then $v(\mathrm{tr}(h)) = -\|w\|$. It follows that H is freely generated by A and B, and that $v(\mathrm{tr}(h)) < 0$ for all $1 \neq h \in H$.

Now let G be a finitely generated fully residually free group, so by Theorem 4.10 G embeds in some ultrapower ${}^*F_2$, which is isomorphic to ${}^*H$, a subgroup of $\mathrm{SL}_2({}^*P)$ (which is identified in an obvious manner with ${}^*SL_2(P)$). Take a finite set of generators for G, and let L be the field generated (over $\mathbb{Q}$) by the entries of these generators and their inverses (viewing G as a subgroup of $\mathrm{SL}_2({}^*P)$). Every element of G can be represented by a word in the generators and their inverses, and by induction on the length of the word, it is easy to see that G embeds in ${}^*H \cap \mathrm{SL}_2(L)$. The valuation v extends to a valuation ${}^*v$ of ${}^*P$, and by restriction we obtain a valuation v_1 of L. Now v_1 inherits from v the property that $v_1(\mathrm{tr}(g)) < 0$ for all $1 \neq g \in G$, and it follows from Lemma 4.7 and the remarks following it that G acts freely on the Λ-tree (X_{v_1}, d), where Λ is the value group of v_1. Also, v_1 is identically zero on $\mathbb{Q}^\times$, and so Λ is finitely generated (see Chapter VII, §10.3, Corollaire 1 in [**9**]).

However, recent progress has improved on Theorem 4.11. This involves the idea of a group admitting exponents from an arbitrary ring, an idea of Lyndon [**36**].

DEFINITION. Let R be a ring. An R-group is a group G together with a mapping $G \times R \to G$, $(g, \alpha) \mapsto g^\alpha$, satisfying the following, for all g, $h \in G$ and all α, $\beta \in R$

(1) $g^1 = g$.
(2) $g^{\alpha+\beta} = g^\alpha g^\beta$.
(3) $g^{\alpha\beta} = (g^\alpha)^\beta$.
(4) $h^{-1}g^\alpha h = (h^{-1}gh)^\alpha$.

Clearly the R-groups, for a given R, form a category. It follows from general results in universal algebra that given a set X, there is a free object in this category on the set X, which we denote by $F(X)^R$. Lyndon established a normal form theorem for free R-groups, where $R = \mathbb{Z}[t_1, \dots, t_n]$, the polynomial ring in finitely many commuting indeterminates. Using this, Pfander [**44**] has shown that $F_2^{\mathbb{Z}[t]}$, the free $\mathbb{Z}[t]$-group on two generators, embeds in $F_2^I/\mathcal{D}$, where I is a countable set and $\mathcal{D}$ is a non-principal ultrafilter on I. He further shows that if G is a finitely generated subgroup of $F_2^{\mathbb{Z}[t]}$, then the Lyndon length function ${}^*L$ on ${}^*F_2$ constructed above, when restricted to G, takes values in a group isomorphic to $\mathbb{Z} \oplus \cdots \oplus \mathbb{Z}$, the direct sum of finitely many copies of $\mathbb{Z}$ with the lexicographic ordering. Using this and the results of Bass mentioned in Example 5 above, he shows that G is finitely presented.

Myasnikov and Kharlampovich have shown that any finitely generated fully residually free group embeds in $F_2^{\mathbb{Z}[t]}$. Their results were presented in their lectures, and we refer to their articles in this volume. They also have an alternative method of showing that finitely generated subgroups of $F_2^{\mathbb{Z}[t]}$ are finitely presented, and it follows that finitely generated fully residually free groups are finitely presented. A proof of this last result has also been announced by Z. Sela.

We note that Myasnikov and Kharlampovich use an extra axiom in their definition of an R-group: if g, h commute, then $(gh)^\alpha = g^\alpha h^\alpha$, for all g, $h \in G$ and $\alpha \in R$. All free R-groups in the sense above satisfy this axiom.

General properties of tree-free groups. One simple observation is that tree-free groups are torsion-free, because hyperbolic elements act as translations on their axes. A more profound result is the following. It was originally proved, in the case $\Lambda = \mathbb{R}$, in an equivalent formulation using Lyndon length functions, by N. Harrison [**31**]. For the general result, see [**13**] and [**55**].

THEOREM 4.12. *Let G be a tree-free group. Then any subgroup which can be generated by at most two elements is either free of rank* 2 *or abelian.*

Of course if the subgroup in 4.12 is abelian, it will be free abelian since it is torsion-free. Before stating the next result on tree-free groups, some definitions are needed.

DEFINITION. A group G is commutative transitive if the following equivalent conditions are satisfied.

(1) Given a, b, $c \in G$ such that $[a, b] = 1$, $[b, c] = 1$ and $b \neq 1$, then $[a, c] = 1$.
(2) Centralizers of non-identity elements of G are abelian.
(3) If M_1, M_2 are maximal abelian subgroups of G and $M_1 \neq M_2$, then $M_1 \cap M_2 = \{1\}$.

The following definition appears in [**41**] and [**30**].

DEFINITION. A group G is called a CSA group if the maximal abelian subgroups of G are malnormal. That is, if M is a maximal abelian subgroup and $g \in G$ is such that $M \cap gMg^{-1} \neq \{1\}$, then $g \in M$.

EXERCISE. Show that the three conditions in the definition of commutative transitive group are indeed equivalent, and that CSA implies commutative transitive.

THEOREM 4.13. *A tree-free group is* CSA *and commutative transitive.*

PROOF. See 1.8 and 1.9 in [**3**]. The exercise (CSA implies commutative transitive) is not needed.

Thus there are the following inclusions of classes of groups:

tree-free groups $\subset$ CSA groups $\subset$ commutative transitive groups.

The right-hand inclusion is strict, as noted in Remark 5 of [**41**] (a simple example is the infinite dihedral group). The left-hand inclusion is also strict. In [**16**] there are examples of word hyperbolic groups of cohomological dimension 2 (so torsion-free) which are not tree-free. These are CSA by Proposition 12 in [**41**].

Using Rips' characterization of finitely generated $\mathbb{R}$-free groups, we can prove the following.

LEMMA 4.14. *An $\mathbb{R}$-free group is locally fully residually free.*

PROOF. We need to show that finitely generated $\mathbb{R}$-free groups are fully residually free. By the argument in the proof of Theorem 6 in [**4**], the free product of two fully residually free groups is fully residually free. By 4.8, finitely generated free abelian groups and the non-exceptional surface groups are fully residually free. The result follows by induction and Rips' Theorem.

By Theorem 4.10, locally fully residually free groups embed in ultrapowers of F_2, so are tree-free. We therefore have further inclusions:

$\mathbb{R}$-free groups $\subset$ locally fully residually free groups $\subset$ tree-free groups.

Moreover the inclusions are strict. For the right-hand inclusion, we use the following example from [**28**]. Let G be the HNN-extension with presentation $\langle t, b, c \mid t(b^{-1}c)t^{-1} = cb\rangle$, a group which is tree-free by the results of Bass (see Example 5 above). Put $x_1 = t$, $x_2 = t^{-1}b^{-1}c$, $x_3 = c^{-1}$ and apply Tietze transformations, to obtain the presentation

$$G = \langle x_1, x_2, x_3 \mid x_1^2x_2^2x_3^2 = 1\rangle$$

so G is an exceptional surface group. It is not residually free because in a free group, the equation $a^2b^2c^2 = 1$ implies that a, b and c commute (see [**37**]). Thus if F is a free group and $\phi: G \to F$ is any homomorphism, $\phi([x_1, x_2]) = 1$, but $[x_1, x_2] \neq 1$ in G. This can be seen by viewing G as the free product of a free group generated by x_1 and x_2 and the cyclic group generated by x_3, amalgamating the cyclic subgroups generated by $x_1^2x_2^2$ and x_3^{-2}.

To see that the left-hand inclusion is strict, note that by the remark after Corollary 4.9 and Rips' Theorem, abelianising a finitely generated $\mathbb{R}$-free group results in a finite direct product of infinite cyclic groups and cyclic groups of order

2. Now $G = \langle x_1, x_2, x_3, x_4 \mid x_1^3 x_2^3 x_3^3 x_4^3 = 1\rangle$ is fully residually free by §7 in [**4**], but is not $\mathbb{R}$-free since its abelianisation contains 3-torsion.

One other point is the following, noted in [**28**]. If G is a finitely generated fully residually free group, then the maximal abelian subgroups of G are finitely generated. For G is Λ-free for some finitely generated Λ, by 4.11. If A is an abelian subgroup, then obviously $\ell([g,h]) = 0$ for all g, $h \in A$, where as usual ℓ means hyperbolic length. By the remarks preceding Theorem 4.5, the action of A is abelian and there is a homomorphism $\tau\colon A \to \Lambda$ such that $\ell(g) = |\tau(g)|$ for all $g \in A$. Since the action is free and without inversions, $\ker(\tau)$ is trivial, hence A embeds in Λ, so is finitely generated. This raises the following questions.

QUESTION 1. If G is a finitely generated tree-free group, is G Λ-free for some finitely generated ordered abelian group Λ?

QUESTION 2. If G is a finitely generated tree-free group, are its abelian subgroups finitely generated?

Before posing a third question, another definition is needed. A *right-ordered group* is a group G with a linear ordering $\le$ on G such that, if g, h and $k \in G$ and $g \le h$, then $gk \le hk$. If, in addition, $g \le h$ implies $kg \le kh$ for all g, h and $k \in G$, then G is called an *ordered group*. This is of course a generalisation of the idea of an ordered abelian group. A group is called right-orderable (resp. orderable) if there exists a linear ordering on it making it into a right-ordered (resp. ordered) group.

Now free groups, in particular F_2, are orderable (see [**43**], Chapter 13, Corollary 2.8), and the ordering on F_2 can be extended to an ordering on any ultraproduct of F_2 as indicated in the discussion of ultraproducts above. It is easy to see that this makes any such ultraproduct into an ordered group. Clearly subgroups of ordered groups are ordered. It follows from Theorem 4.10 that locally fully residually free groups are orderable. This leads to the following question.

QUESTION 3. Is every tree-free group orderable, or at least right-orderable?

References

1. R. C. Alperin and H. Bass, *Length functions of group actions on Λ-trees*, Combinatorial Group Theory and Topology (S. M. Gersten and J. R. Stallings, eds.), University Press, Princeton; Ann. of Math. Stud. **111** (1987), 265–378.
2. R. C. Alperin and K. N. Moss, *Complete trees for groups with a real-valued length function*, J. London Math. Soc. (2) **31** (1985), 55–68.
3. H. Bass, *Group actions on non-archimedean trees*, Arboreal Group Theory (R. C. Alperin, ed.), Math. Sci. Res. Inst. Publ. **19**, Springer-Verlag, New York, 1991, pp. 69–131.
4. B. Baumslag, *Residually free groups*, Proc. London Math. Soc. (3) **17** (1967), 402–418.
5. G. Baumslag, *On generalized free products*, Math. Z. **78** (1962), 423–438.
6. J. L. Bell and A. B. Slomson, *Models and ultraproducts*, North-Holland, Amsterdam, 1971.
7. M. Bestvina and M. Feighn, *Stable actions of groups on real trees*, Invent. Math. **121** (1995), 287–321.
8. ———, *Outer limits*, preprint.
9. N. Bourbaki, *Éléments de mathématique* XXX*, algèbre commutative*, Hermann, Paris, 1964.
10. K. S. Brown, *Trees, valuations and the Bieri-Neumann-Strebel invariant*, Invent. Math. **90** (1987), 479–504.
11. I. M. Chiswell, *Abstract length functions in groups*, Math. Proc. Cambridge Philos. Soc. **80** (1976), 451–463.
12. ———, *The Bass-Serre theorem revisited*, J. Pure Appl. Algebra **15** (1979), 117–123.
13. ———, *Harrison's theorem for Λ-trees*, Quart. J. Math Oxford (2) **45** (1994), 1–12.

14. ———, *Introduction to Λ-trees*, Semigroups, Formal Languages and Groups (J. Fountain, ed.), Dordrecht, Boston, London, Kluwer, 1995.
15. ———, *Properly discontinuous actions on Λ-trees*, Proc. Edinburgh Math. Soc. **37** (1994), 423–444.
16. ———, *Some examples of groups having no non-trivial action on a Λ-tree*, Mathematika **42** (1995), 214–219.
17. D. E. Cohen, *Combinatorial group theory: a topological approach*, London Math. Soc. Student Text **14**, University Press, Cambridge, 1989.
18. M. M. Cohen and M. Lustig, *Very small group actions on ℝ-trees and Dehn twist automorphisms*, Topology **34** (1995), 575–617.
19. M. Coornaert, T. Delzant, and A. Papadopoulos, *Géométrie et théorie des groupes*, Lecture Notes in Math. **1441**, Springer, Berlin, Heidelberg, 1990.
20. M. Culler and J. Morgan, *Group actions on ℝ-trees*, Proc. London Math. Soc.(3) **55** (1987), 571–604.
21. M. Culler and K. Vogtmann, *Moduli of graphs and automorphisms of free groups*, Invent. Math. **84** (1986), 91–119.
22. ———, *The boundary of outer space in rank two*, Arboreal Group Theory (Berkeley, CA, 1988) (R.C. Alperin, ed.), Math. Sci. Res. Inst. Publ. **10**, Springer, New York, 1991, pp. 189–230.
23. M. J. Dunwoody, *Groups acting on protrees* (to appear).
24. D. Gaboriau and G. Levitt, *The rank of actions on ℝ-trees*, Ann. Sci. École Norm. Sup. (4) **28** (1995), 549–570.
25. D. Gaboriau, G. Levitt, and F. Paulin, *Pseudogroups of isometries of ℝ and Rips' theorem on free actions on ℝ-trees*, Israel. J. Math. **87** (1994), 403–428.
26. A. M. Gaglione and D. Spellman, *Does Lyndon's length function imply the universal theory of free groups*, The Mathematical Legacy of Wilhelm Magnus: Groups, Geometry and Special Functions (Brooklyn, NY, 1992), Contemp. Math., vol. 169, Amer. Math. Soc., Providence, RI, 1994, pp. 277–281.
27. ———, *Every 'universally free' group is tree free*, Group Theory (Granville, Ohio 1992), World Sci. Publishing, River Edge, NJ, 1993, pp. 149–154.
28. ———, *Generalisations of free groups: some questions*, Comm. Algebra **22** (1994), 3159–3169.
29. E. Ghys and P. de la Harpe, *Sur les groupes hyperboliques d'après Mikhael Gromov*, Birkhäuser, Boston, 1990.
30. D. Gildenhuys, O. Kharlampovich, and A. G. Myasnikov, *CSA groups and separated free constructions*, Bull. Austral. Math. Soc. **52** (1995), 63–84.
31. N. Harrison, *Real length functions in groups*, Trans. Amer. Math. Soc. **174** (1972), 77–106.
32. W. Imrich, *On metric properties of tree-like spaces*, Beiträge zur Graphentheorie und deren Anwendungen (Sektion MARÖK der Technischen Hochschule Ilmenau, ed.), Oberhof, DDR, 1977, pp. 129–156.
33. S. Krstic and K. Vogtmann, *Equivariant outer space and automorphisms of free-by-finite groups*, Comment. Math. Helv **68** (1993), 216–262.
34. G. Levitt, *Constructing free actions on ℝ-trees*, Duke Math. J. **69** (1993), 615–633.
35. R. C. Lyndon, *Length functions in groups*, Math. Scand. **12** (1963), 209–234.
36. ———, *Groups with parametric exponents*, Trans. Amer. Math. Soc. **96** (1960), 518–533.
37. ———, *The equation $a^2b^2 = c^2$ in free groups*, Michigan Math. J. **6** (1959), 155–164.
38. J. W. Morgan, *Λ-trees and their applications*, Bull. Amer. Math. Soc. **26** (1992), 87–112.
39. J. W. Morgan and P. B. Shalen, *Valuations, trees and degenerations of hyperbolic structures: I.*, Annals of Math. (2) **122** (1985), 398–476.
40. ———, *Free actions of surface groups on ℝ-trees*, Topology **30** (1991), 143–154.
41. A. G. Myasnikov and V. N. Remeslennikov, *Exponential groups 2: extensions of centralizers and tensor completions of CSA-groups*, Internat. J. Algebra Comput. (to appear).
42. H. Neumann, *Varieties of groups*, Springer, Berlin, Heidelberg, New York, 1967.
43. D. S. Passman, *The algebraic structure of group rings*, John Wiley, New York, London, Sydney, Tokyo, 1977.
44. P. H. Pfander, *Finitely generated subgroups of the free $\mathbb{Z}[t]$ group on two generators*, London Math. Soc. Lecture Notes, Cambridge University Press (to appear).
45. V. N. Remeslennikov, *∃-free groups*, Siberian Math. J. **30** (1989), 998–1001.

46. ______, $\exists$-*free groups as groups with length function*, Ukrainian Math. J. **44** (1992), 733–738.
47. J.-P. Serre, *Trees*, Springer-Verlag, New York, 1980.
48. P. B. Shalen, *Dendrology of groups: an introduction*, Essays in Group Theory (S. M. Gersten, ed.), Math. Sci. Res. Inst. Publ. **8**, Springer-Verlag, New York, 1987, pp. 265–319.
49. ______, *Dendrology and its applications*, Group Theory from a Geometrical Viewpoint (E. Ghys, A. Haefliger, and A. Verjovsky, eds.), World Scientific, Singapore, New Jersey, London, Hong Kong, 1991, pp. 543–616.
50. H. Short (ed.), *Notes on word hyperbolic groups*, Group Theory from a Geometrical Viewpoint (E. Ghys, A. Haefliger, and A. Verjovsky, eds.), World Scientific, Singapore, New Jersey, London, Hong Kong, 1991, pp. 3–63.
51. R. Skora, *Deformations of length functions in groups*, preprint.
52. J. R. Stallings, *Foldings of G-trees*, Arboreal Group Theory (R. C. Alperin, ed.), Math. Sci. Res. Inst. Publ. **19**, Springer-Verlag, New York, 1991, pp. 355–368.
53. M. Steiner, *Gluing data and group actions on* Λ-*trees*, preprint.
54. J. Tits, *A "theorem of Lie-Kolchin" for trees*, Contribution to algebra: A collection of papers dedicated to Ellis Kolchin, Academic Press, New York, 1977, pp. 377–388.
55. M. Urbański and L. Q. Zamboni, *On free actions on* Λ-*trees*, Math. Proc. Cambridge. Philos. Soc. **113** (1993), 535–542.
56. L. van den Dries and A. Wilkie, *Gromov's theorem on groups of polynomial growth and elementary logic*, J. Algebra **89** (1984), 349–374.
57. C. T. C. Wall and G. P. Scott, *Topological methods in group theory*, Homological Group Theory (C. T. C. Wall, ed.), London Math. Soc. Lecture Notes **36**, University Press, Cambridge, 1979, pp. 137–203.
58. T. White, *Fixed points of finite groups of free group automorphisms*, Proc. Amer. Math. Soc. **118** (1993), 681–688.
59. H. Zieschang, E. Vogt, and H.-D. Coldewey, *Surfaces and planar discontinuous groups*, Lecture Notes in Mathematics **835**, Springer, Berlin, Heidelberg, New York, 1980.

School of Mathematical Sciences, Queen Mary and Westfield College, University of London, Mile End Road, London E1 4NS, England.

E-mail address: I.M.Chiswell@qmw.ac.uk

Centre de Recherches Mathématiques
CRM Proceedings and Lecture Notes
Volume 17, 1999

Introduction to Hyperbolic and Automatic Groups

S. M. Gersten

1. Introduction

These notes comprise the revised and edited text of two hour lectures I delivered at the CRM Summer School on Groups held at Banff in August 1996. I made no attempt at completeness, but tried to introduce the subject to novices from the attractive drawings of M. Escher, and I made guesses from the list of speakers about what aspects of the subject I should introduce to make my lectures useful to them. Thus I began at the beginning, and to quote G. Baumslag, ended very near the beginning. I have included a short bibliography of source material on the subject along with some of my personal comments scattered in the text about what I have found useful from these references, each of which is good in its own way for an aspect of the subject. The reader should consult these for a more extensive bibliography.[1]

1.1. We start with a comparison of two lithographs of M. Escher.

Figure 1 is part of a tiling of the Euclidean plane, which we imagine as continued in all directions, and Figure 2 is a beautiful tesselation of the Poincaré unit disc model of the hyperbolic plane by white tiles representing angels and black tiles representing devils. An important feature of the second is that all white tiles are mutually congruent as are all black tiles; of course this is not true for the Euclidean metric, but holds for the Poincaré metric $ds^2 = 4(dx^2 + dy^2)/(1 - r^2)^2$, where $r^2 = x^2 + y^2$.[2] One feature distinguishing the figures is immediate. There is no natural boundary associated to the Euclidean tiling, whereas the boundary of the disc limits the hyperbolic tiling. This circle is not part of the hyperbolic plane, which in this model consists of points in the interior of the disc, but nevertheless it is apparent from the model.

1991 *Mathematics Subject Classification.* 20F05, 20F32, 57M07.
Partially supported by NSF grant DMS-9500769.
This is the final form of the paper.

[1]Theorems, examples, or exercises are indicated with an asterisk * if the proof is omitted or if the sketch given is not in keeping with the otherwise elementary character of these notes. These theorems so indicated can be looked up in the references, and the starred exercises or examples ignored, if the terms are unfamiliar. I have tried to make this exposition intelligible to a reader who has completed a year's graduate study or has a good undergraduate background in math.

[2]The irritating factor 4 is present to guarantee that the curvature is -1.

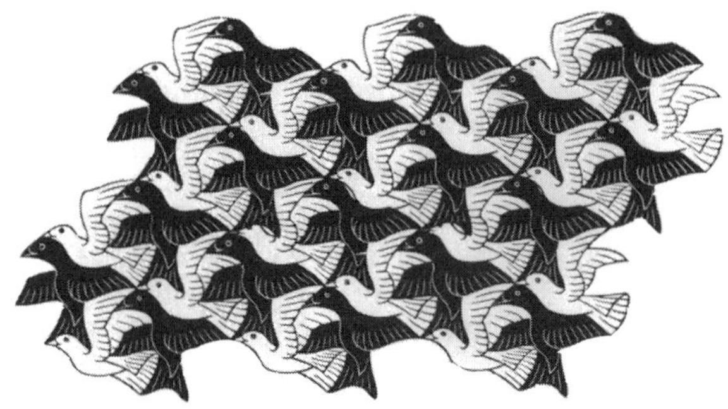

FIGURE 1. Euclidean Escher picture. (M. C. Escher "Regular division of the plane with birds" ©1998 Cordon Art B. V., Baarn, Holland. All rights reserved.)

FIGURE 2. Hyperbolic Escher picture. (M. C. Escher "Circle limit IV" ©1998 Cordon Art B. V., Baarn, Holland. All rights reserved.)

A second less immediate feature distinguishing them is the so-called isoperimetric inequality. If we imagine circles of radius R drawn from a fixed center, then the number of tiles in the Euclidean drawing in the interior of the circle of radius R for large R is between C_1R^2 and C_2R^2, where $C_1 < C_2$ are positive constants. In the hyperbolic figure, when we calculate distances in the Poincaré metric, we find that the number of black and white tiles in the interior of a circle of radius R is bounded by CR, for some constant $C > 0$. We say that the Euclidean plane satisfies the quadratic isoperimetric inequality, whereas the hyperbolic plane satisfies the linear isoperimetric inequality. I shall have much more to say about isoperimetric inequalities later.

The third difference I want to point out is more subtle but is crucial for our departure into hyperbolic groups, the property of thin triangles for the hyperbolic plane. For this, and in general for all calculations, it is more convenient to use the conformally equivalent model of the upper half plane $\mathbb{H}$. This is the set of points

(x, y) in the Cartesian plane with $y > 0$ equipped with the Riemannian metric

$$ds_{\mathbb{H}}^2 = \frac{ds_{\mathbb{E}}^2}{y^2} = \frac{dx^2 + dy^2}{y^2}, \tag{1.1}$$

where $ds_{\mathbb{E}}$ is the Euclidean metric. We denote the distance between points P, Q in $\mathbb{H}$ by $d_{\mathbb{H}}(P, Q)$ and we recall that this is determined by minimizing the lengths $\int_\gamma ds$ of all piecewise smooth paths γ from P to Q in our metric $ds_{\mathbb{H}}$.

Geodesics for this geometry are vertical lines together with arcs of circles orthogonal to the x-axis. Every pair of points in $\mathbb{H}$ is joined by a unique geodesic segment and every geodesic segment can be extended indefinitely in both directions.

A geodesic triangle is one whose sides are segments of geodesics. There are also the so-called ideal geodesic triangles, whose vertices in the disc model lie on the limit circle, and in the model $\mathbb{H}$ lie on the x-axis, where the geodesics are considered as extended arbitrarily far in both directions. One also says that one end point of a vertical geodesic lies "at the point of infinity". This corresponds to a point on the limit circle in the disc model under the conformal equivalence, but is not visible in $\mathbb{H}$. An important property is that all ideal triangles are equivalent under the isometry group of $\mathbb{H}$.[3]

The property we need is the following

1.2. THEOREM ("δ-thin triangles"). *There is a number $\delta > 0$ so that for all geodesic triangles in $\mathbb{H}$ with vertices, say, A, B, and C, and all points P on side AB, there exists a point Q on at least one of the sides AC or CB so that the distance $d_{\mathbb{H}}(P, Q) \leq \delta$. The optimal value of δ is $\ln(1 + \sqrt{2})$.*

PROOF. Let P be on the side AB of the geodesic triangle ABC. First note that by moving the point C toward infinity away from P along the extension of AC, the distance of P to the opposite two sides is only increased. Similarly, we can move the point A to infinity and then move the point B, so we may assume the triangle ABC is ideal. Then using the fact that all ideal triangles are equivalent, we may assume that $A = (0, 0)$, $B = (2, 0)$, and C is the point at infinity, so P is on the circle of radius 1 centered at $(1, 0)$. Symmetry considerations then show that it suffices to prove that $d_{\mathbb{H}}(P, Q) \geq d_{\mathbb{H}}(P_1, Q_1)$ in Figure 3, where $P = (1, 1)$; here PQ and P_1Q_1 are segments of geodesics.

But by (1.1), the homothety $z \mapsto kz$ induces an isometry of $\mathbb{H}$, where $k > 0$ and where z is the complex coordinate of a point. Since PQ and P_1Q_1 are on circles centered at $(0, 0)$, there is a unique value of k for which the homothety takes the second arc onto the arc P_2Q, where P_2 is a point of the arc PQ. It follows that $d_{\mathbb{H}}(P_1, Q_1) = d_{\mathbb{H}}(P_2, Q) \leq d_{\mathbb{H}}(P, Q)$.

Thus all geodesic triangles in $\mathbb{H}$ are $\delta = d_{\mathbb{H}}(P, Q)$-thin. It now becomes a problem of integration to calculate δ.

1.2.1. EXERCISE. Carry out the calculation. (Hint. Parametrize the arc PQ on the circle of radius $\sqrt{2}$ by $x = \sqrt{2}\cos\theta$, $y = \sqrt{2}\sin\theta$, $\pi/4 \leq \theta \leq \pi/2$, and calculate $\delta = \int_{\pi/4}^{\pi/2} \csc\theta \, d\theta = \ln(1 + \sqrt{2})$.)

[3] This follows from the facts (1) that linear fractional transformations $x \mapsto (ax + b)(cx + d)^{-1}$ act transitively on sets consisting of three distinct points of $\mathbb{R} \cup \{\infty\}$, where a, b, c and d are real numbers and $ad - bc = 1$, and (2) that each such linear fractional transformation acts as an isometry of $\mathbb{H}$.

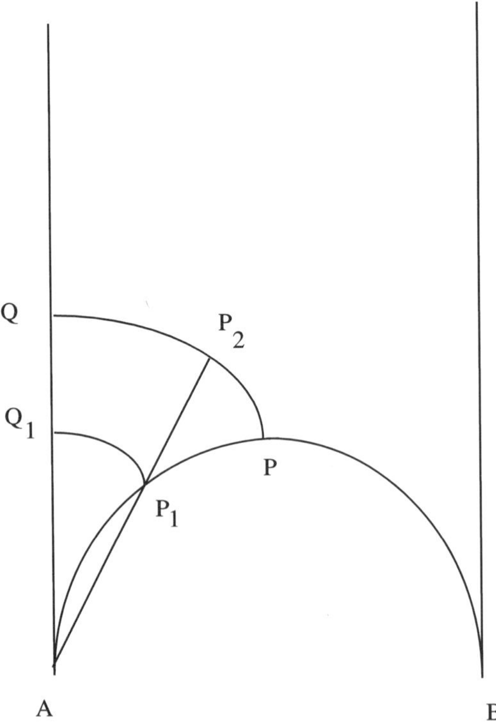

FIGURE 3. Thin triangles.

2. Geodesic Metric Spaces

A metric space (X, d) is called geodesic if for all pairs of points x, y in X there is an isometric imbedding $f\colon \big[0, d(x, y)\big] \to X$ taking the end points of the interval to x and y. The image of f is called a geodesic segment connecting these points, or more simply, a geodesic.

2.1. A complete smooth connected Riemannian manifold with its path metric obtained by minimizing lengths $\int_\gamma ds$ of all piecewise smooth paths γ between two points is an example of a geodesic metric space. Note that in contrast with the terminology of Riemannian geometry, our geodesics are minimal geodesics, in the sense that their arc length is the distance between the end points. For example the equator of the round unit sphere is a geodesic in the sense of Riemannian geometry, but a segment on it of length at most π is a geodesic segment in the sense we are using.

2.2. A second example of a geodesic metric space is a Cayley graph of a finitely generated group G. Suppose that A is a finite set of generators for G, in the sense that every element of G can be written as a finite product of elements of A and their inverses. We define a graph $\Gamma_{G,A}$ whose vertex set is G and whose edges are all triples (g, a, g'), where $g, g' \in G$, $a \in A$, and $ga = g'$. This edge is considered to originate at the vertex g and terminate at the vertex g'. The group G acts on the left by left translation on vertices and $x(g, a, g') = (xg, a, xg')$ for the action of the

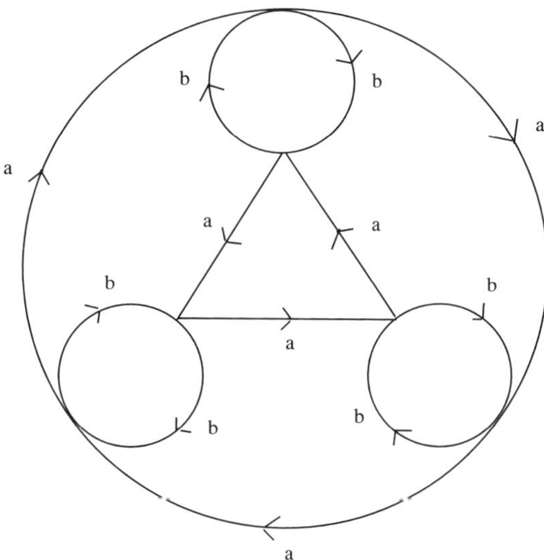

FIGURE 4. Cayley graph of S_3.

group element x on the edge (g, a, g'). That this makes sense is a consequence of the associative law $x(ga) = (xg)a$.

We can realize this graph as a 1-dimensional CW complex where for each abstract edge (g, a, g') we attach a copy of the unit interval with end points 0,1 identified with vertices g, g' respectively. We shall not distinguish between the graph and its geometric realization in this exposition.

The graph $\Gamma_{G,A}$ admits a natural metric, which we now describe. If $g, g' \in G$, we set $d(g, g')$ to be the minimum length of an edge-path connecting these vertices, where each edge of $\Gamma_{G,A}$ has length 1. This is the same as the minimum length of a word w in the elements of A and their inverses such that $gw = g'$ when the product is evaluated in G, and hence it is called the *word metric* on G for generators A. This metric is extended to all pairs of points of the geometric realization by taking the path metric induced by requiring left action of the group to be an isometric action on the edges and requiring each edge to be isometric to the unit interval in $\mathbb{R}$. Thus each point is at distance at most $\frac{1}{2}$ from a vertex, and the path metric agrees with the metric already defined on pairs of vertices. With this metric the graph $\Gamma_{G,A}$ becomes a geodesic metric space, and the left translations by elements of G become isometries.

Cayley graphs are objects of great beauty. The example above in Figure 4 is a Cayley graph for the symmetric group S_3 and a generating set consisting of a 2-cycle and a 3-cycle.

Another example is the free group F_n with free basis $A = \{x_1, x_2, \ldots, x_n\}$. The Cayley graph $\Gamma_{F_n,A}$ is a tree, where each vertex has valence $2n$. The fact that it is a tree is a reflection of the uniqueness of the expression of an element of F_n as a reduced word in the generators and their inverses.

2.3. EXERCISE. Show conversely that if a Cayley graph $\Gamma_{G,A}$ is a tree, then the group G is free with free basis A.

2.4. Edge-circuits in $\Gamma_{G,A}$ also have significance. Each oriented edge is equipped with a label in $A \cup A^{-1}$. The positive orientation of (g, a, g') has label a whereas if we read the edge with the opposite orientation the label is a^{-1}. The label of an edge-circuit $\mathfrak{c}$ is thus a word w in $A \cup A^{-1}$ which multiplies the initial vertex of $\mathfrak{c}$ to itself, and hence w evaluates in G to give the identity element of G. Thus labels of edge-circuits give relations among the generators A and their inverses, and furthermore, given a word w in $A \cup A^{-1}$ and a vertex v, there is a unique edge-circuit starting at v with label w. Thus up to choice of initial point, which can be taken to be the identity element of G, there is a $1-1$ correspondence between edge-circuits of $\Gamma_{G,A}$ and relations among the generators $A \cup A^{-1}$.

2.5. The Cayley graph and the word metric depend on the choice of generators. A useful strategy in the search for group theoretic invariants is to define a property of a Cayley graph and then elucidate how the property behaves with respect to change of generators. This will be our strategy in defining the notion of hyperbolicity of a group and in defining isoperimetric functions.

3. δ-Hyperbolicity

We say that a geodesic metric space (X, d) is δ-hyperbolic, where $\delta \geq 0$, if for every geodesic triangle ABC and point P on the segment AB there exists a point Q in the union of the sides AC and BC so that $d(P, Q) \leq \delta$.

3.1. Examples.

3.1.1. The hyperbolic plane $\mathbb{H}$ is $\ln(1 + \sqrt{2})$-hyperbolic.

3.1.2. Every simplicial tree (*i.e.* a 1-dimensional contractible simplicial complex with the path metric, where each edge has length 1) is 0-hyperbolic.

3.1.3. If a geodesic metric space has bounded diameter D, then it is D-hyperbolic.

3.1.4. The Euclidean plane $\mathbb{E}$ is not δ-hyperbolic for any choice of δ. For $\mathbb{E}$ admits a scaling μ_k for each real number $k > 0$ which multiplies all distances by k. Thus if ABC is any nondegenerate triangle and P is on side AB and at distance $d > 0$ from the union of the sides AC and BC, then $\mu_k(P)$ is at distance kd from the sides $\mu_k(A)\mu_k(C) \cup \mu_k(B)\mu_k(C)$ of the triangle $\mu_k(A)\mu_k(B)\mu_k(C)$. Letting $k \to \infty$ establishes the assertion.

3.2. We define an $\mathbb{R}$-tree to be a 0-hyperbolic geodesic metric space. Such spaces will be considered in depth in I. Chiswell's lectures in this volume.

3.3. We say a finitely generated group G is hyperbolic[4] if for some finite set of generators A of G the Cayley graph $\Gamma_{G,A}$ is δ-hyperbolic. According to the strategy proposed at the end of 2.5 to show that hyperbolicity is a group theoretic property we must show that the Cayley graph $\Gamma_{G,A'}$ for any other finite set of generators A' of G is δ'-hyperbolic for some $\delta' \geq 0$. This will be deduced as a consequence of properties of quasi-geodesics, which we take up next.

3.4. Definition. A (λ, ϵ) quasi-isometric mapping $f\colon (X, d) \to (X', d')$ between two metric spaces (X, d), (X', d') is a (discontinuous) function f satisfying

---

[4]The terms negatively curved and word hyperbolic and even Gromov hyperbolic have been used to describe the same notion, but we shall say simply hyperbolic.

the inequalities

$$\frac{1}{\lambda}d'\big(f(x), f(y)\big) - \epsilon \leq d(x,y) \leq \lambda d'\big(f(x), f(y)\big) + \epsilon$$

for all $x, y \in X$; here $\lambda > 0$ and $\epsilon \geq 0$. Note that a $(\lambda, 0)$ quasi-isometric mapping is just a bilipschitz mapping with lipschitz constant $\lambda > 0$ and is *a fortiori* continuous. It is the possibility that $\varepsilon > 0$ in this notion that allows for discontinuity of the map f. One can think of a quasi-isometric mapping as the view a farsighted person with astigmatism has of the world. The curvature of the lens of the eye accounts for the multiplicative distortion factor λ in distances while the additive constant ϵ has the interpretation that distances closer than ϵ are not resolved clearly.

3.5. A (λ, ϵ) quasi-geodesic mapping of an interval $[a, b] \subset \mathbb{R}$ into (X, d) is called a (λ, ϵ) quasi-geodesic. If we let $a = 0$, $b = +\infty$, then it is called a quasi-geodesic ray. Often one considers a family of (λ, ϵ) quasi-geodesics simultaneously; one calls members of this family quasi-geodesics if the constants λ, ϵ are understood and are the same for all members of the family.

3.6. Two metric spaces (X, d) and (X', d') are quasi-isometric if there are numbers $\lambda > 0$, $\epsilon \geq 0$, and $C \geq 0$, and (λ, ϵ) quasi-isometric mappings $f: (X, d) \to (X', d')$ and $f': (X', d') \to (X, d)$ so that both compositions $f \circ f'$ and $f' \circ f$ are within C of the appropriate identity map. Thus $d\big(f'(f(x)), x\big) \leq C$ for all $x \in X$, and similarly for the other composition.

3.7. EXAMPLE. Let A, A' be two finite sets of generators for the same group G and consider the word metrics d, d' on G for these sets of generators. Now each element $a \in A$ can be written as a word in the generators A' and their inverses, and similarly each $a' \in A'$ can be written as a word in the generators A and their inverses. Let M be the maximum lengths of these words that arise in rewriting the members of each of the two generating sets in terms of the other generators. Then it is easy to see that the identity map $G \to G$ is an $(M, 0)$ quasi-isometry. If we recall that G is the vertex set for each of the two Cayley graphs $\Gamma_{G,A}$, $\Gamma_{G,A'}$ and that every point of a Cayley graph is at distance at most $\frac{1}{2}$ from a vertex, we see that the identity map of G extends to quasi-isometric maps $f: \Gamma_{G,A} \to \Gamma_{G,A'}$ and $f': \Gamma_{G,A'} \to \Gamma_{G,A}$ whose compositions both ways are within finite distance of the appropriate identity maps.

3.7.1. EXERCISE. Verify that the numbers $\lambda = M$, $\epsilon = M + 1$, and $C = (M^2 + 1)/2$ work for the maps f, f' just defined.

Observe that the finiteness of the generating sets A, A' was used in the definition of the number M. This is the only place so far where finite generation was used.

It follows from this discussion that

3.8. PROPOSITION. *Any two Cayley graphs for the same finitely generated group are quasi-isometric geodesic metric spaces.*

3.9. DEFINITION. If A and B are two subsets of the metric space (X, d), we say that A and B are at finite Hausdorff distance if there is a constant $H > 0$ so that for every $a \in A$ there is a point $b \in B$ with $d(a, b) \leq H$ and for every $b \in B$ there is a point $a \in A$ with $d(a, b) \leq H$. Thus B is contained in the H-neighborhood of the set A and A is contained in the H-neighborhood of B.

The basic result connecting quasi-isometry and hyperbolicity is the following.

3.10. THEOREM* (Quasi-geodesics are close to geodesics). *Let X be a δ-hyperbolic geodesic metric space. Then there is a function $H(\lambda, \epsilon) > 0$ so that for any two points $x, y \in X$, any (λ, ϵ) quasi-geodesic $f\colon [a,b] \to X$ with end points mapped to x, y, and any geodesic segment γ with the same end points x, y one has the image of f and the image of γ are at finite Hausdorff distance at most $H(\lambda, \epsilon)$ from each other.*

The result is nontrivial. I find the proof given in the Berkeley notes [**Bkly**], which is based on exponential divergence of geodesics in a hyperbolic metric space, very accessible. An easier argument for the special case of a complete Riemannian manifold with negative sectional curvatures bounded above by $-\kappa < 0$ appears in [**Wordproc**]. My understanding is that the result was first observed for the hyperbolic plane by Marston Morse in the 20's, although the terminology of quasi-geodesics was only used by G. Mostow and others after the 60's.

3.10.1. EXAMPLE. The following example shows one way in which the conclusion of Theorem 3.10 can fail in a nonhyperbolic group. Consider the group $G = \mathbb{Z}^2$ with generating set $A = \{a = (1,0), b = (0,1)\}$. If n is a positive number, then it is easy to verify that all geodesics in $\Gamma_{G,A}$ from $(0,0)$ to (n,n) are of the form words in a and b which involve exactly n a's and exactly n b's (and no a^{-1} nor b^{-1}). Two extreme geodesics are given by $a^n b^n$ and $b^n a^n$. The distance between the vertex $(n,0)$ on the first and $(0,n)$ on the second is $2n$, which can be arbitrarily large. Thus geodesics segments with the same end points can be at arbitrarily large Hausdorff distance in this example.

If we anticipate Section 8 on regular languages below, the set of all geodesic words in this example is given by the regular expression $L = \{a,b\}^* + \{A,b\}^* + \{A,B\}^* + \{a,B\}^*$, where $A = a^{-1} = (-1,0)$ and $B = b^{-1} = (0,-1)$. Thus the set of all geodesic words in this example is a regular language, but the group is not hyperbolic.

3.10.2. REMARK. An interesting converse to Theorem 3.10 was discovered by P. Papasoglu [**Papa**] in the special case of Cayley graphs. He showed that if G is a finitely generated group with finite set A of generators and if there is a number $H \geq 0$ so that all pairs of geodesic segments in $\Gamma_{G,A}$ with the same endpoints are at Hausdorff distance at most H, then G is hyperbolic.

Theorem 3.10 will now be applied to show that the property of a finitely generated group being hyperbolic is independent of the particular finite set of generators.

3.11. THEOREM (quasi-isometry invariance of hyperbolicity). *Let (X, d) and (X', d') be quasi-isometric geodesic metric spaces. If (X, d) is δ-hyperbolic, then there exists $\delta' \geq 0$ so that (X', d') is δ'-hyperbolic.*

PROOF. Suppose $f\colon X \to X'$ and $f'\colon X' \to X$ are both (λ, ϵ) quasi-isometric maps whose compositions both ways are within $C \geq 0$ of the appropriate identity map, and we assume that (X, d) is δ-hyperbolic. Let Δ' be a geodesic triangle in X'. Then $f'(\Delta')$ is a quasi-geodesic triangle in X whose sides are (λ, ε) quasi-geodesics. It follows from Theorem 3.10 that there is a geodesic triangle Δ in X with the same vertices whose sides are each within Hausdorff distance $H = H(\lambda, \varepsilon)$ of the corresponding sides of $f'(\Delta')$. It follows that $f'(\Delta')$ is $(2H + \delta)$-thin, where we can talk about thinness of quasi-geodesic triangles in an obvious sense. Now apply f to get the quasi-geodesic triangle $f\big(f'(\Delta')\big)$ in X, which one sees is $\big(\lambda(2H+) + \varepsilon\big)$-thin from the way quasi-isometric maps distort distances. But points on the sides of the triangle $f(f'(\Delta'))$ are at distance at most C from the corresponding points on the original triangle Δ. It follows that Δ is $; = \big(2C + \lambda(2H + \ ) + \varepsilon\big)$-thin. Thus all geodesic triangles of X' are ;-thin for this value of ;, and the result is proved.

3.12. It follows from Theorem 3.11 that *if the Cayley graph $\Gamma_{G,A}$ is δ-hyperbolic, then there exists $\delta' \geq 0$ so that the Cayley graph $\Gamma_{G,A'}$ is δ'-hyperbolic*; here G is a finitely generated group and A and A' are finite sets of generators for G.

It follows that hyperbolicity is a property of groups. In particular although it is defined using one finite system of generators, the property carries over to any other finite system of generators (for a different value of δ).

At the moment, we know from examples and exercises that finite groups and finitely generated free groups are hyperbolic. We now prepare to state a result which provides many more examples of hyperbolic groups.

3.13. Definition. An action of the group G on the topological space X is a mapping $G \times X \to X$, denoted $(g, x) \mapsto gx$, so that X is a G-set (recall that this means that $1x = x$ for all $x \in X$ and $g'(gx) = (g'g)x$ for all $g, g' \in G$, $x \in X$) and so that for each $g \in G$ the map $x \mapsto gx$ is a homeomorphism of X onto itself. We are considering G here to have the discrete topology.

The action is called *properly discontinuous* if for each compact subset K of X the collection of groups elements $\{g \in G \mid gK \cap K \neq \emptyset\}$ is finite. A special case of a properly discontinuous action where all point stabilizers in G are trivial (so $gx = x$ for $g \in G$, $x \in X$, implies $g = 1$) is called a *free action*.

The action is called *cocompact* if the orbit space $G\backslash X$ is compact.

The metric space (X, d) is called *proper* if all balls of finite radius are precompact, in the sense that their closures are compact. In this case X is locally compact.

In many of the examples we shall consider, X will be a metric space and the action will often be by isometries, so that the homeomorphisms of X given by left translation $x \mapsto gx$, $g \in G$, $x \in X$, are isometries of X. We say such actions are *isometric*.

3.14. Exercises.

3.14.1. If $\pi\colon Y \to X$ is a regular covering map with covering group G, then the action of G on Y by deck transformations is a free action. If X is compact, then the action is cocompact.

3.14.2. The left action of the finitely generated group on its Cayley graph $\Gamma_{G,A}$ is free and cocompact and group elements act by isometries. In addition $\Gamma_{G,A}$ is a proper metric space.

3.14.3. Let M be a closed (and hence compact) connected Riemannian manifold and let $\widetilde{M}$ be its universal cover equipped with the pull-back Riemannian metric from the covering projection $\pi\colon \widetilde{M} \to M$. If G is the fundamental group of M, then the action of G on $\widetilde{M}$ by deck transformations is free, isometric, and cocompact, and $\widetilde{M}$ is a proper geodesic metric space.

3.15. Theorem*. *Suppose that the group G acts properly discontinuously and cocompactly by isometries on the proper geodesic metric space (X, d). Then G is finitely generated, and G with any word metric and (X, d) are quasi-isometric metric spaces.*

A very readable account of this result is given in M. Troyanov's article in the Swiss notes [**Swiss**]. The result has some important corollaries, which we state as exercises.

3.16. Exercises.

3.16.1. Any Cayley graph $\Gamma_{G,A}$ of the finitely generated group G for finite set A of generators is quasi-isometric to G with the word metric. In fact we already observed this by direct calculation in 3.7.1.

3.16.2. If H is a subgroup of finite index of the group G, then H is finitely generated iff G is finitely generated, and in this case H and G are quasi-isometric for their word metrics. (Hint. The first assertion can be established by covering space theory. As for the second, the action of H on $\Gamma_{G,A}$ obtained by restricting the left action of G to H satisfies the hypotheses of Theorem 3.15.)

3.16.3. Let N be a *finite* normal subgroup of the finitely generated group G. Then the groups G and G/N are quasi-isometric for their word metrics. (Hint. Consider the action of G on any Cayley graph of G/N by left translation by its associated coset modulo N.)

3.16.4. If M is a closed connected Riemannian manifold and $\widetilde{M}$ is the universal cover of M with the pull-back Riemannian metric, then $\widetilde{M}$ is quasi-isometric to the fundamental group of M. This result is due to Švarc and independently to Milnor (who did not use the language of quasi-isometry to state it). They used it to relate the growth of finitely generated groups to the growth of volumes of balls in the universal cover.

3.17. We can now give the promised examples of hyperbolic groups. Note first that hyperbolic n-space $\mathbb{H}^n$ is defined as $\{(x_1, x_2, \ldots, x_n) \in \mathbb{R}^n \mid x_n > 0\}$ with the Riemannian metric

$$ds^n_{\mathbb{H}^n} = \frac{ds^n_{\mathbb{E}^n}}{x_n^2} = \frac{x_1^2 + x_2^2 + \cdots + x_n^2}{x_n^2},$$

where $ds_{\mathbb{E}^n}$ is the Euclidean metric. It has the property that every geodesic triangle is contained in a totally geodesic isometric copy of the hyperbolic plane, so $\mathbb{H}^n$ is δ-hyperbolic for $\delta = \ln(1+\sqrt{2})$. A (closed) hyperbolic n-manifold can be defined as the quotient of $\mathbb{H}^n$ by a cocompact properly discontinuous subgroup of isometries acting freely on $\mathbb{H}^n$.[5]

3.17.1. If G is a subgroup of the group of isometries of $\mathbb{H}^n$ which acts properly discontinuously and cocompactly on $\mathbb{H}^n$, then G is (finitely generated and) hyperbolic. This is immediate from Theorem 3.15 and remarks above.

It follows that the symmetry group of the Escher diagram in Figure 2 is a hyperbolic group.

3.17.2*. If G is the fundamental group of a closed Riemannian manifold M all of whose sectional curvatures are strictly negative, then G is hyperbolic. This is an application of the comparison theorems of differential geometry. There is an upper bound $-k$, for some $k > 0$, on the sectional curvatures of $\widetilde{M}$, by the compactness of M. Then the comparison theorem of Toponogov says that geodesic triangles in $\widetilde{M}$ are at least as thin as those in the plane with metric of constant curvature $-k$.

[5]This is equivalent to the usual definition in Riemannian geometry of a hyperbolic manifold as a closed Riemannian manifold which admits a metric with constant sectional curvatures -1.

However in this latter space triangles are scaled copies of triangles in $\mathbb{H}$ and hence are uniformly thin.

3.17.3. If H is a subgroup of finite index of the finitely generated group G, then H is hyperbolic iff G is hyperbolic.

3.17.4. Virtually finitely generated free groups are hyperbolic. Here, if $\mathcal{P}$ is a property of groups (so $\mathcal{P}$ is a proper class of groups closed under isomorphism), then a group is called virtually $\mathcal{P}$ if it has a subgroup of finite index which is in $\mathcal{P}$.

It is amusing to note that finite groups can be characterized as virtually trivial groups.

3.17.5*. Fundamental groups of closed orientable surfaces of genus at least 2 or closed nonorientable surfaces of (nonorientable) genus at least 3 are hyperbolic.[6] This follows from the uniformization theorem, which in this context states that these surfaces admit hyperbolic structures. A more combinatorial proof of this result will be given in §6.

Triangle groups*. Let $2 < m < n < p$ be integers such that $1/m+1/n+1/p < 1$. Then one shows there exists a geodesic triangle Δ in $\mathbb{H}$ with angles π/m,π/n, and π/p. Consider the pattern generated by reflecting the triangle in its sides, and the resulting figure in all its sides, and so on. It follows from a nontrivial theorem of Poincaré's that the resulting pattern tesselates $\mathbb{H}$ by congruent copies of Δ with disjoint interiors. A readable account of this result is given in [**Beard**]. The full isometry group of this tesselation is a properly discontinuous cocompact subgroup of isometries of $\mathbb{H}$. It is generated by reflections in the sides of the original triangle Δ and is called the triangle group $\Delta(m, n, p)$. It is hyperbolic by Theorem 3.15.

There are other important examples of hyperbolic groups, including the [2]CAT(-1)-groups and uniform lattices in rank 1 Lie groups, which time considerations did not permit me to discuss. I shall discuss the important examples of small cancellation groups in § 6 below.

4. The Boundary of a Hyperbolic Group

This notion never came up in the other lectures at the conference, so I shall be brief.

4.1. Let $\Gamma_{G,A}$ be the Cayley graph for the finitely generated group G with finite set A of generators. Assume that $\Gamma_{G,A}$ is δ-hyperbolic. If R and R' are geodesic rays (*i.e.* $(1,0)$ quasi-geodesic rays, see 3.5), then we write $R \sim R'$ if they are in a Hausdorff neighborhood of each other. This is clearly an equivalence relation and the boundary ∂G of G is the set of equivalence classes of geodesic rays under this equivalence relation.

4.1.1. Every quasi-geodesic ray is in a Hausdorff neighborhood of a geodesic ray.

Hence ∂G could have been defined in an equivalent manner as the set of equivalence classes of quasi-geodesic rays, where two quasi-geodesic rays are equivalent if they are in Hausdorff neighborhoods of each other.

4.1.2. Consider the disc model for the hyperbolic plane. Show that every geodesic ray determines a unique point on the limit circle and that two rays are equivalent iff they determine the same point.

---

[6]An orientable surface has genus g if it is the connected sum of n tori and a nonorientable surface has genus g if it is the connected sum of g projective planes.

4.1.3*. Given any point v in $\Gamma_{G,A}$ and geodesic ray R there exists a geodesic ray R' so that $R'(0) = v$ and $R \sim R'$. In fact one considers geodesic segments $vR(t)$ and one shows, using the Ascoli theorem, that these subconverge on compact subintervals of $[0, \infty)$ to a geodesic ray. For full details consult [**Stras**] Chapter 2.

4.2. There is a natural topology on ∂G for which it is a compact finite dimensional metrizable space [**Bndry**], called the visual topology. To describe the topology, it is convenient to choose a base point v. Every ray is equivalent to one starting at v, so choose a ray R with $R(0) = v$. If $t, \varepsilon > 0$ let $N_\varepsilon(R,t)$ be the set of equivalence classes of rays S starting at v such that $d\big(R(t), S(t)\big) \leq \varepsilon$. These sets are a subbasis for the topology on ∂G. In fact, one only needs to use one value of $\varepsilon = 10\delta$ to define this topology [**Bndry**].

4.2.1. If G and G' are quasi-isometric hyperbolic groups, show that their boundaries ∂G and $\partial G'$ are homeomorphic.

4.2.2. Show that the isometric action of G on $\Gamma_{G,A}$ defines an action on rays preserving the equivalence relation $\sim$. Show that this induces an action of G on ∂G where each group element acts as a homeomorphism of ∂G.

The boundary is one of the reasons that topologists are so interested in hyperbolic groups. N. Benakli showed in her thesis that the Menger and Sierpiński curves occur naturally as boundaries of hyperbolic groups. The characterization of hyperbolic groups which have the circle as boundary, achieved independently by Gabai, Casson-Jungreis, and Tukia, led to the solution of a classical conjecture of Seifert's, that closed irreducible 3-manifolds containing normal infinite cyclic subgroups in their fundamental groups are Seifert fibred. One of the outstanding problems of 3-dimensional topology is the conjecture that a closed irreducible 3-manifold with an infinite hyperbolic fundamental group admits a Riemannian metric of constant negative curvature. As a result of the work of Bestvina and Mess, it is known that the boundary of such a group is homeomorphic to the 2-sphere.

5. Finite Presentability of Hyperbolic Groups

Let G be a finitely generated group with Cayley graph $\Gamma_{G,A}$ for finite set A of generators and suppose that $\Gamma_{G,A}$ is δ-hyperbolic. Suppose that γ, γ' are geodesics beginning at 1 and ending at g, γ' respectively and suppose further that $g' = ga$, where $a \in A$. Then we have a geodesic triangle Δ where one side is of length 1 and the other two sides are γ and γ'. It is convenient to extend the domains of definition of γ, γ' to all nonnegative reals by insisting they be the constant maps after they reach their endpoints. This convention will be followed without further mention.

5.1. Lemma. *We have* $d\big(\gamma(t), \gamma'(t)\big) \leq 2(\delta + 1)$ *for all* $t \geq 0$.

Proof. If P is in the image of γ, then P is at distance at most δ from the union of the other two sides of Δ. If P is unlucky enough to be within δ of the edge-side, then it is at distance at most $\delta + 1$ from the end point of γ'. Thus in any case P is at most $\delta + 1$ from the image of γ'.

Thus we see that for all $0 \leq t \leq d(1, g)$ there exists an $0 \leq s \leq d(1, g')$ so that $d\big(\gamma(t), \gamma'(s)\big) \leq \delta + 1$. But γ and γ' are geodesics in this range of values of their domains, so $|t - s| \leq \delta + 1$ by the triangle inequality. Now using the fact that γ' is geodesic on $\big[0, d(1, g')\big]$ we see that $d\big(\gamma'(t), \gamma'(s)\big) \leq |t - s| \leq \delta + 1$. Hence by the triangle inequality we have $d\big(\gamma(t), \gamma'(t)\big) \leq 2(\delta + 1)$, completing the proof.

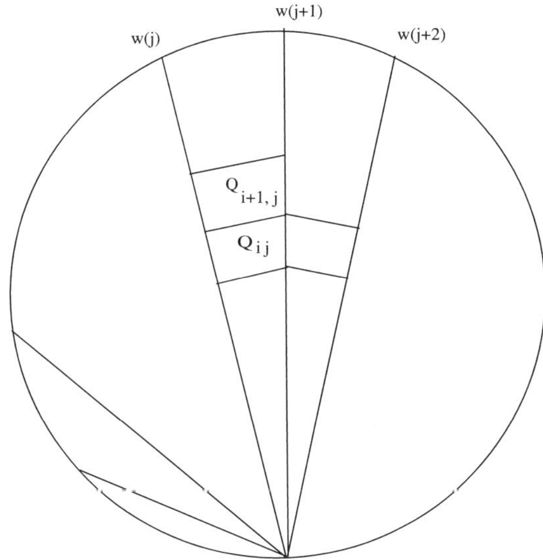

FIGURE 5. Filling the circuit w.

5.2. Now we mark off "integer points" $\gamma(i)$, $\gamma'(i)$ on the geodesics γ, γ', $i \in \mathbb{N}$. We obtain for each i a quadrilateral Q_i, where two sides are geodesics chosen to join $\gamma(i)$ to $\gamma'(i)$ and $\gamma(i+1)$ to $\gamma'(i+1)$ and the remaining two sides are edges of $\Gamma_{G,A}$ joining $\gamma(i)$ to $\gamma(i+1)$ and joining $\gamma'(i)$ to $\gamma'(i+1)$ (if i lies beyond the range where γ is geodesic, interpret this to be the constant path at the vertex $\gamma(i)$). The boundary label of Q_i is a word in the generators $A \cup A^{-1}$ of length at most $2(2(\delta+1)+1) = 4\delta+6$ which evaluates to 1 in G by the discussion of 2.4. We let $\mathcal{R}$ be the set of all words in $A \cup A^{-1}$ of length at most $4\delta+6$ which evaluate to 1 in G. Note that $\mathcal{R}$ is a finite set of words.

5.3. THEOREM. *$\mathcal{P} = \langle A \mid \mathcal{R}\rangle$ is a finite presentation for G.*

PROOF. Let w be an edge-circuit in $\Gamma_{G,A}$ with $w(0) = 1$. We "cone" w from the base point 1 by geodesics, so, precisely, we choose geodesics γ_j from 1 to $w(j)$. For each i we have the geodesic quadrilateral Q_{ij} with sides joining vertices $\gamma_j(i)$, $\gamma_j(i+1)$, $\gamma_{j+1}(i+1)$, $\gamma_{j+1}(i)$ and back to $\gamma_j(i)$. These quadrilaterals fit together to form a disc diagram filling w, as is illustrated in Figure 5.

But the label of Q_{ij} is one of the defining relators of $\mathcal{P}$, and it follows that the label of w, which is a relation among the generators, is a consequence of the relators $\mathcal{R}$. Thus $\mathcal{P}$ is a finite presentation for G.

5.4. Much more than finite presentability is true for hyperbolic groups. The known finiteness conditions follow from Rips's theorem*: *if G is a hyperbolic group, then there is a contractible finite dimensional locally finite simplicial complex P and a properly discontinuous cocompact simplicial action of G on P.* As a consequence, if G is torsion-free, the action of G on P is free and the orbit map $P \to G\backslash P$ is a covering map, so $G\backslash P$ is a compact Eilenberg-MacLane space of type $K(G,1)$. All three references [**Bkly**], [**Stras**], and [**Swiss**] give readable accounts of Rips's theorem.

5.5. In general, the minimum number of relators needed in a filling of a relation w of a finite presentation $\mathcal{P}$ is called the area of w and is written $\text{Area}_{\mathcal{P}}(w)$. It is equal to the minimum number of factors in an expression for w as a product of conjugates of defining relators and their inverses. The equivalence of these two expressions for the area of w is a consequence of a lemma of van Kampen's and was treated in S. Ivanov's lectures at the conference.

5.6. One can estimate the area of w in the proof of Theorem 5.3, using the fact that the length of the geodesic γ_j is equal to $d(1, w(j)) \leq j$. The result is $\text{Area}_{\mathcal{P}}(w) \leq \sum_{i=1}^{\ell(w)} i \leq C\ell(w)^2$, where $\ell(w)$ is the length of the circuit w and where C is a constant. In the terminology of §6, this shows that hyperbolic groups satisfy the quadratic isoperimetric inequality. In fact, much more is true.

5.7. THEOREM*. *Let $\mathcal{P} = \langle A \mid \mathcal{R} \rangle$ be any finite presentation for a hyperbolic group G. Then there exists $K > 0$ so that for all edge-circuits w in $\Gamma_{G,A}$ one has*

$$\text{Area}_{\mathcal{P}}(w) \leq K\ell(w). \tag{5.1}$$

Conversely, if a finitely presented group G satisfies the linear isoperimetric inequality (5.7.1) *for some finite presentation $\mathcal{P}$ of G, then G is hyperbolic.*

Both assertions in Theorem 5.7 are nontrivial. I have found the treatment in [**Bkly**] to be useful. It is also worth remarking that Ol'shanskii gave an accessible treatment of a strengthening of the converse assertion, due originally to Gromov, that a finitely presented group satisfying a subquadratic isoperimetric inequality is hyperbolic [**Olshan**]. Papasoglu and Bowditch have also treated the question. I have treated the question of a cohomological interpretation of the linear isoperimetric inequality, and the final word has not yet been written on the matter.

6. Small Cancellation Groups

In this section I shall show that the classical small cancellation groups of type $C'(1/6)$ are hyperbolic, making use of a combinatorial argument I gave in the same volume where Gromov's fundamental paper [**Gromov**] appeared.

6.1. We consider in this section combinatorial cell complexes of dimension 2. Technically, this is just a finite 2-dimensional CW-complex X in which each attaching map f of a 2-cell is a map of a circle subdivided into a finite number of intervals such that f restricted to the interior of each such interval is a homeomorphism onto an edge of the 1-skeleton. Thus f itself need not be $1-1$, but no nondegenerate subinterval of the circle is mapped to a point. A 2-cell $\mathcal{F}$ of X is called a face, and the number of intervals into which the domain of the attaching map of $\mathcal{F}$ is subdivided as above is denoted $n(\mathcal{F})$.

6.1.1. EXAMPLE. Let $\mathcal{P} = \langle x_1, x_2, \ldots, x_p \mid R_1, R_2, \ldots, R_q \rangle$ be a finite presentation. Thus each R_i is a word in the free group $F(x_1, x_2, \ldots, x_p)$ with free basis $x_1, x_2, \ldots, x_p$. The 2-complex $X_{\mathcal{P}}$ canonically associated with $\mathcal{P}$ has one 0-cell, p 1-cells e_i in $1-1$ correspondence with the generators x_i each with a preferred orientation or label x_i, and q 2-cells $\mathcal{F}_j$ in $1-1$ correspondence with the relators R_j, where the attaching map of $\mathcal{F}_j$ spells out the word R_j as a map of the circle, subdivided into $n(\mathcal{F}_j) = \ell(R_j)$ segments, to the 1-skeleton; here $\ell(R_j)$ is the length of the free word R_j.

6.2. A *corner* α of a 2-cell $\mathcal{F}$ of X is simply one of the subdivision points of the domain of the attaching map of $\mathcal{F}$. It is better to think of α as an angle between two adjacent sides of a polygon with $n(\mathcal{F})$ sides, but this is just a convenient visualization. A *weight* function on X is an assignment of real numbers $w(\alpha)$ to the corners α of 2-cells of X. In this exposition we demand that $w(\alpha) > 0$ but there are situations we shall not discuss here where even this is too restrictive a condition.

6.2.1. Example.

If X is a regular n-gon in the Euclidean plane, $n \geq 3$, then X has one 2-cell $\mathcal{F}$ with $n(\mathcal{F}) = n$, and one can assign to each corner the Euclidean angle $(n-2)\pi/n$. Note that in this case the sum of the weights attached to the corners of $\mathcal{F}$ is $(n-2)\pi$, the sum of the interior angles of a convex Euclidean n-gon.

The following result is completely elementary, although the proof involves somewhat tedious calculations.

6.3. PROPOSITION. *Suppose that the 2-complex X is a topological 2-dimensional disc equipped with a weight function w such that*

(1) *there is an $\varepsilon > 0$ so that for each interior vertex v of D the sum of the weights of corners of 2-cells incident at v is at least $2\pi + \varepsilon$, and*
(2) *for each 2-cell $\mathcal{F}$ of D the sum of the weights of corners of $\mathcal{F}$ is at most $(n(\mathcal{F}) - 2)\pi$.*

Then $F \leq (1 + 2\pi/\varepsilon)E_\infty$, where F is the number of 2-cells of D and E_∞ is the number of edges on the boundary of D.

PROOF. We denote by V, V_{int}, V_∞, E, and E_{int} the number of vertices, number of interior vertices, number of boundary vertices, number of geometric edges, and number of geometric edges in the interior of D, respectively. Note that $E = E_{\text{int}} + E_\infty$, $V = V_{\text{int}} + V_\infty$, $E_\infty = V_\infty$, and $V - E + F = 1$, where the last two equalities come from the calculation of the Euler characteristics of the circle and disc as 0 and 1, respectively.

Let $S = \sum_\alpha w(\alpha)$, where the sum is over all corners α of all 2-cells of D. By (1) we have

$$(2\pi + \varepsilon)V_{\text{int}} = \sum_{v \in \text{int}(D)^{(0)}} (2\pi + \varepsilon) \leq S,$$

where the sum in the middle is over vertices in the interior of D. By (2) we have

$$S \leq \sum_{\mathcal{F}} (n(\mathcal{F}) - 2)\pi = \pi \sum_{\mathcal{F}} n(\mathcal{F}) - 2\pi F,$$

where the sums are over all 2-cells $\mathcal{F}$ of D. Combining the two displayed inequalities gives

$$(2\pi + \varepsilon)V_{\text{int}} \leq \pi \sum_{\mathcal{F}} n(\mathcal{F}) - 2\pi \mathcal{F}. \tag{6.1}$$

Now substitute the equalities

$$\sum_{\mathcal{F}} n(\mathcal{F}) = 2E_{\text{int}} + E_\infty = 2E - E_\infty$$

and $V_{\text{int}} = V - E_\infty$ into (6.1) and manipulate to obtain

$$\varepsilon V \leq (\pi + \varepsilon)E_\infty - 2\pi(V - E + F) = (\pi + \varepsilon)E_\infty - 2\pi \leq (\pi + \varepsilon)E_\infty,$$

or

$$V \leq \frac{\varepsilon + \pi}{\varepsilon} E_\infty. \tag{6.2}$$

Next note that our assumption that $w(\alpha) > 0$ for all corners α and (2) imply that $n(\mathcal{F}) \geq 3$ for all 2-cells $\mathcal{F}$ of D. Thus $3F \leq \sum_{\mathcal{F}} n(\mathcal{F}) = 2E - E_\infty$, or

$$(3F + E_\infty)/2 \leq E. \tag{6.3}$$

Use Euler's formula $E - F + 1 = V$ again, replace E by (6.3) and replace V by (6.4) to get

$$(3F + E_\infty)/2 - F + 1 \leq \frac{\varepsilon + \pi}{\varepsilon} E_\infty,$$

and solve to get $F \leq (1 + 2\pi/\varepsilon)E_\infty - 2 \leq (1 + 2\pi/\varepsilon)E_\infty$, which establishes the result.

6.4. To apply the preceding result we need the notion of a reduced disc diagram in our 2-complex X. First, a disc diagram is a combinatorial map $f: D \to X$, where D is a combinatorial cell complex whose underlying space is the 2-dimensional disc and where f is a combinatorial map, in the sense that f restricted to each open cell of D is a homeomorphism *onto* an open cell of the same dimension of X. The map f is called *reduced* if it is never the case that there are 2 2-cells $\mathcal{F}$, $\mathcal{F}'$ in D with an oriented edge e in common so that f maps the boundaries of these cells, read as edge-loops in $X^{(1)}$ beginning with the letter $f(e)$, to the same word in the oriented edges of X.

6.5. We can now introduce the small cancellation condition on X. Let $f: D \to X$ be a reduced disc diagram. Let $\overline{D}$ be obtained from D by removing all *interior* vertices of valence 2 (so the boundary of D is unchanged). Let $\overline{\mathcal{F}}$ be the face of $\overline{D}$ corresponding to the face $\mathcal{F}$ of D and let $n(\overline{\mathcal{F}})$ be its number of corners (this is just $n(\mathcal{F})$ reduced by the number of vertices in the boundary of $\mathcal{F}$ which are of valence 2 in the interior of D). Give the corners of $\overline{\mathcal{F}}$ the weights of a regular Euclidean $n(\bar{\mathcal{F}})$-gon (*cf.* 6.2.1). This puts a weight function w on the corners of $\overline{D}$. We say that X *satisfies the small cancellation condition* if this weight function w satisfies conditions (1) and (2) for one fixed ε and all choices of f.[7]

6.6. EXERCISE. Show that if $\mathcal{P}$ is a finite presentation satisfying the small cancellation condition $C'(1/6)$ introduced in S. Ivanov's lectures,[8] then the 2-complex $X_\mathcal{P}$ satisfies the small cancellation condition. The number ε in this case can be taken to be $\pi/7$; this minimum is achieved with 3 7-gons meeting at an interior vertex of $\overline{D}$.

6.7. THEOREM. *Let $\mathcal{P}$ be a finite presentation satisfying the small cancellation condition $C'(1/6)$. Then every reduced disc diagram $f: D \to X_\mathcal{P}$ satisfies the linear isoperimetric inequality*

$$F \leq 15E_\infty.$$

[7] I have called this the condition of negative curvature in print. However that term is getting somewhat overworked and suggests some connection with geometry, whereas our weights have no *a priori* connection with geometry.

[8] This means each relator of $\mathcal{P}$ is cyclically reduced, and the length of every piece of a relator is *strictly less* than one-sixth the length of that relator, *cf.* [**L-S**]. Recall that a *piece* is the label of an arc in the interior of a reduced 2-faced disc diagram. The convention of [**L-S**] also requires that $\mathcal{P}$ contain all cyclic conjugates of every relator and its inverse.

PROOF. This is immediate from Proposition 6.3 and Exercise 6.6.

6.8. COROLLARY. *If a group G admits a finite presentation satisfying the small cancellation condition $C'(1/6)$, then G is hyperbolic.*

PROOF. This is immediate from Theorems 6.7 and 5.7.

6.9. Show that the standard presentations of fundamental groups of closed orientable surfaces of genus at least 2 satisfy the small cancellation condition $C'(1/6)$. This gives another proof (*cf.* 3.17.5) that these groups are hyperbolic.

Let us also mention the following result proved by the same method.

6.10. THEOREM. *Suppose that the finite 2-complex X is such that there is a number $\varepsilon > 0$ such that every reduced disc diagram $f: D \to X$ has a weight satisfying (1) and (2) for this number ε. Then the fundamental group of X is hyperbolic.* □

6.11. EXAMPLE. Consider the presentation $\mathcal{P} = \langle x, y, z \mid x^2y^2z^2 \rangle$ for the fundamental group of the nonorientable surface of genus 3. This does not satisfy condition $C'(1/6)$ since each piece is exactly one-sixth the length of the defining relator. However an assignment of weight $\pi/2$ to all corners in a reduced disc diagram satisfies (1) and (2), as the reader should check, and it follows that the group of $\mathcal{P}$ is hyperbolic.

6.12. EXAMPLE*. If $\mathcal{P}$ is a finite presentation satisfying the small cancellation condition $C'(1/6)$ and such that that no relator of $\mathcal{P}$ is a proper power in the free group on the generators of $\mathcal{P}$, then the group G of the presentation is torsion-free. For let $\mathcal{Q}$ be the presentation obtained from $\mathcal{P}$ be choosing precisely one representative from each cyclic conjugacy class of relator or its inverse and let $X = X_{\mathcal{Q}}$. Note that the group of $\mathcal{Q}$ is also G. The small cancellation condition implies that $\pi_2(X)$ is generated as a $\pi_1(X)$-module by classes of 2-faced *spherical* diagrams, where a spherical diagram is a combinatorial map of a cell structure on the 2-sphere into X. But the condition that no relator is a proper power means all 2-faced spherical diagrams are null-homotopic, so it follows that $\pi_2(X) = 0$. Thus, X is aspherical in the sense that it is a $K(G, 1)$. It follows from the P. A. Smith theorem (which states that a group G with a finite dimensional $K(G, 1)$ is torsion free) that G is torsion-free.

6.13. A theorem of Rips's [**Rips**] states that given any finitely presented group G there is a short exact sequence of groups

$$1 \to N \to E \to G \to 1$$

where E has a finite presentation satisfying the small cancellation condition $C'(1/6)$ and where N is a normal subgroup which is finitely generated as a group. In Rips's construction the group N is never finitely presented except in trivial cases, so this gives examples of finitely generated subgroups of hyperbolic groups which are not finitely presented. I showed in [**Gerst2**] that all finitely presented subgroups of $C'(1/6)$-groups G (and of certain other hyperbolic groups G like those of cohomological dimension 2) were hyperbolic. Recently N. Brady [**Brady**] has given an example of a finitely presented subgroup H of a hyperbolic group G of cohomological dimension 3 where H is not hyperbolic. Beyond these results there is almost nothing known about the subgroup structure of general hyperbolic groups.

7. Isoperimetric Functions and Dehn Functions

In this section I want to digress on isoperimetric inequalities, both as a bridge between hyperbolic and automatic groups, and because I consider these to offer an important new outlook on finitely presented groups.

7.1. Let $\mathcal{P}$ be a finite presentation with $G = G(\mathcal{P})$ the group determined by the presentation. An isoperimetric function for $\mathcal{P}$ is a function $h: \mathbb{N} \to \mathbb{N}$ such that for all relations w of length at most n one has $\mathrm{Area}_{\mathcal{P}}(w) \leq h(n)$. There exists a minimal isoperimetric function for $\mathcal{P}$ called the *Dehn function* $f_{\mathcal{P}}$, where

$$f_{\mathcal{P}}(n) = \max\{\mathrm{Area}_{\mathcal{P}}(w) \mid \ell(w) \leq n,\ w \equiv 1 \text{ in } G\}.$$

Note that there are only a finite number of relations w among the generators of length at most n, so $f_{\mathcal{P}}(n)$ is a well-defined integer.

7.2. To state in what sense the Dehn function is an invariant, it is necessary to introduce an equivalence relation on functions from $\mathbb{N}$ to $\mathbb{N}$. If $f, g: \mathbb{N} \to \mathbb{N}$ we write $f \preceq g$ if there exist $A, B, C, D, E > 0$ so that $f(n) \leq Ag(Bn + C) + Dn + E$ for all $n \in \mathbb{N}$. We write $f \sim g$ if both $f \preceq g$ and $g \preceq f$. It is clear that $\sim$ is an equivalence relation.

7.3. If $\mathcal{P}$ and $\mathcal{Q}$ are finite presentations for isomorphic groups, then $f_{\mathcal{P}} \sim f_{\mathcal{Q}}$, so their Dehn functions are equivalent. I established this result in [**Gerst**], and it was generalized in [**Alon**] to show that if $\mathcal{P}$ and $\mathcal{Q}$ are finite presentations whose Cayley graphs are quasi-isometric, then their Dehn functions are equivalent.

7.4. Show that if $d, d' \geq 1$ and if the functions $x^d \sim x^{d'}$, then $d = d'$. Thus it makes sense to say that the Dehn function is polynomial, polynomial of degree d, exponential, recursive, etc. It is not too difficult to show that a finite presentation has a solvable word problem iff its Dehn function is recursive.

7.5. The Dehn function "landscape".

7.5.1. It follows from the theorem of Gromov-Ol'shanskii [**Olshan**] that a subquadratic isoperimetric function implies the group is hyperbolic. I showed in [**Gerst2**] that if the Cayley graph is not a tree, then the Dehn function grows at least linearly. Thus, for finite presentations of hyperbolic groups whose Cayley graphs are not trees, the Dehn function is equivalent to x, the identity function. There is no Dehn function equivalent to x^α where $1 < \alpha < 2$.

7.5.2. "The quadratic zoo". The following groups are known to have isoperimetric functions equivalent to x^2 (so for their Dehn functions $f_{\mathcal{P}}$ we have $f_{\mathcal{P}} \preceq x^2$): automatic groups, CAT(0)-groups, and the $(2n+1)$-dimensional integral Heisenberg groups for $n \geq 2$. I shall prove that automatic groups have quadratic isoperimetric functions in §9 below (a theorem of Thurston's).

The integral $(2n+1)$-dimensional Heisenberg group is the group of integral uppertriangular $(n+2)$-by-$(n+2)$ matrices with 1's on the main diagonal and otherwise the only other nonzero entries are in the first row and last column. It was conjectured by Thurston that for $n \geq 2$ these groups had quadratic Dehn functions, and this has recently been established by D. Allcock.

Thurston also conjectured that $\mathrm{Sl}_n(\mathbb{Z})$ for $n \geq 4$ has a quadratic Dehn function, but to my knowledge this is still open.

A CAT(0)-group is a group which acts properly discontinuously and cocompactly on a proper geodesic metric space (X, d) which satisfies the CAT(0) inequality. The CAT(0) inequality means the following. Let ABC be a geodesic triangle in X. Form a comparison triangle $A'B'C'$ in the Euclidean plane $\mathbb{E}$ whose respective side lengths are equal to those of ABC, so $d(A, B) = d_{\mathbb{E}}(A', B')$, etc. Let P be a point on the side BC and let P' be the corresponding point on $B'C'$, so $d(B, P) = d_{\mathbb{E}}(B', P')$. Then it should be the case that $d(A, P) \leq d_{\mathbb{E}}(A', P')$.

I call this a zoo, because I am unable to see any pattern in this bestiary of groups. It would be striking if there existed a reasonable characterization of groups with quadratic Dehn functions, which was more enlightening that saying that they have quadratic Dehn functions.

7.5.3. There is no example known of a group whose Dehn function is equivalent to x^α, where $2 < \alpha < 3$ and it is an interesting open question whether such a group can exist.

7.5.4. The 3 dimensional integral Heisenberg group has Dehn function equivalent to x^3. This is established in [**Wordproc**] and I give a different proof in [**Gerst**].

7.5.5. Exactly which real numbers $\alpha > 2$ are such that x^α is equivalent to the Dehn function of a finitely presented group is an open question at the time of writing. A step in this direction was taken by M. Bridson, who showed that there are infinitely many fractions $\alpha > 3$ so that there exist finite presentations with Dehn functions equivalent to x^α.

It is shown in [**Dehn**] that there is a dense set S of real numbers $\alpha > 3$ so that there exist finite presentations with Dehn functions equivalent to x^α with $\alpha \in S$. This extends earlier results due to M. Bridson on fractional power Dehn functions. Thus there are no "gaps" in the power landscape for $\alpha > 3$.

7.5.6. There are many examples of groups with exponential isoperimetric functions. Among them are $\mathrm{Sl}_3(\mathbb{Z})$ and the Baumslag-Solitar groups $B_{p,q}$ for $1 < p < q$, where $B_{p,q} = \langle x, y \mid yx^p = x^q y\rangle$. These results are proved in [**Wordproc**] and I gave a different proof for $B_{p,q}$ in [**Gerst**].

7.5.7. I showed in [**Gerst**] that the Dehn function for the 1-relator presentation $\langle x, y \mid x^{(x^y)} = x^2\rangle$ grows faster than any *iterated* exponential. Here $x^y = y^{-1}xy$. This is the world record growth so far for 1-relator groups. It can be shown that Ackermann's function, which is recursive but not primitive recursive, is an upper bound for the equivalence class of Dehn functions of all 1-relator groups, but my example has much slower growth than Ackermann's function. It is quite possible that my example is the fastest and indeed this has been announced to me in a private communication. This would make an interesting complement to Magnus's result that 1-relator groups have a solvable word problem.

7.5.8. It has been shown that every recursive function appears as a lower bound for the Dehn function of some finite presentation with a solvable word problem [**BMS**] Corollary 18. Thus the equivalence classes of Dehn functions of finite presentations with solvable word problems have no recursive upper bound under the relation $\preceq$.

8. Automata and Cannon's Theorem

In this section I introduce finite state automata and regular languages and state the theorem of Cannon.

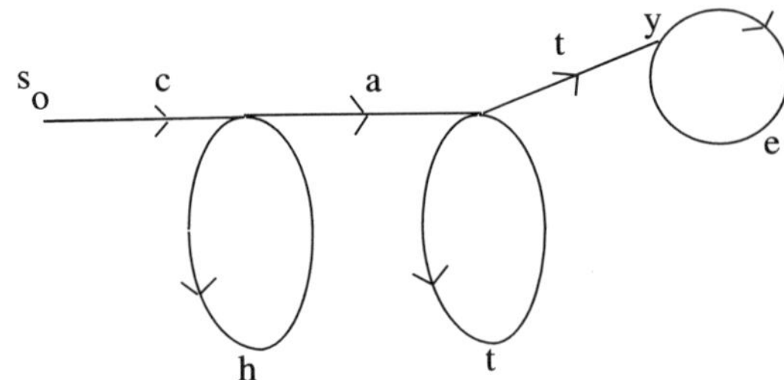

FIGURE 6. A finite state automaton.

8.1. Let A be a finite set and let A^* be the free monoid on the set A. Thus A^* consists of all words (*i.e.* finite concatenations of letters) of the alphabet A including the empty word, which I denote by "1". The monoid operation is concatenation of words. A subset L of A^* is called a language, and, in general, if L is a language, then L^* is by definition the submonoid generated by L. If L and L' are languages, then $L + L'$ denotes their union $L \cup L'$, and LL' is the set of words ww' where $w \in L$ and $w' \in L'$. In case L consists of a single letter $a \in A$, it is traditional to denote L^* by a^* rather than by the cumbersome but correct $\{a\}^*$.

Of particular interest are the regular languages, which are those recognized by finite state automata, which we now define.

8.2. A finite state automaton (FSA) is a 5-tuple $\mathcal{A} = (\Gamma, s_0, A, \lambda, Y)$ where

(1) Γ is a finite directed graph, with vertex set $V(\Gamma)$ and edge set $E(\Gamma)$; the vertices are called "states" and the directed edges are called "transitions";
(2) s_0 is a distinguished vertex of Γ called the "start state";
(3) A is a finite alphabet;
(4) $\lambda\colon E(\Gamma) \to A$ is a function which we think of as labelling the edges of Γ by letters of the alphabet A; and
(5) $Y \subset V(\Gamma)$ is a set of states, possibly empty, called "accept states".

8.2.1. A simple example may help fix notions. Consider the directed graph shown below in Figure 6.

The alphabet $A = \{a, c, e, h, t\}$, $Y = \{y\}$.

8.3. The language $L(\mathcal{A})$ recognized by the FSA $\mathcal{A} = \{\Gamma, s_0, A, \lambda, Y\}$ is by definition the set of labels $\lambda(p)$ of all directed paths p of Γ which begin at the start state s_0 and end at any state of Y; here $\lambda(p)$ is the concatenation of the labels of the edges of p in order. A sublanguage L of A^* which is of the form $L(\mathcal{A})$ for some FSA $\mathcal{A}$ with alphabet A is called a *regular language*.

For example, in the FSA of Figure 6, we have

$$L(\mathcal{A}) = ch^*att^*e^* = \{ch^m at^{n+1}e^p \mid m, n, p \geq 0\}.$$

8.4. Before stating the theorem of Cannon, it is necessary to say some words about generators. Previously we used a set of group theoretic generators to define the edges of Cayley graph, but even there we needed inverses to generate enough paths to connect vertices. For example, if we take an infinite cyclic group $\langle t\rangle$ with generating set $\{t\}$, then there is no edge-path beginning at the vertex t^2, ending at the vertex t, and labelled by positive powers of t. However in discussing languages, we are discussing subsets of free monoids which only involve positive expressions in the free generators.

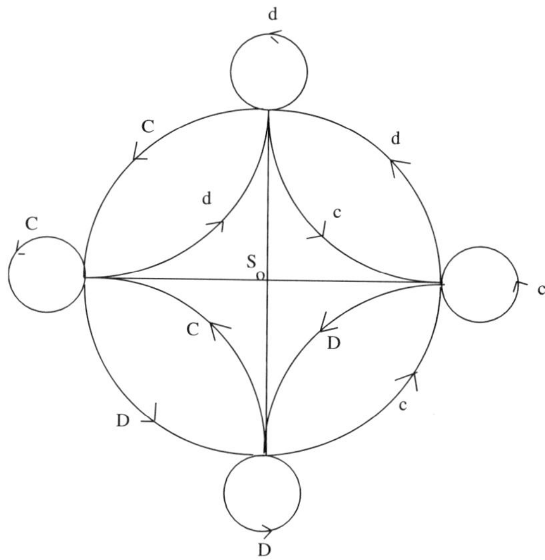

FIGURE 7. Automaton generating all reduced words.

To remedy the problem, we let X be a finite set of group theoretic generators for the finitely generated group G and let $A = X \cup X^{-1}$, a finite set of semigroup generators for G. Let A^* be the free monoid on the set A and let $\mu\colon A^* \to G$ be the evaluation map, which takes a word and evaluates it using the product operation in G. The map μ is surjective precisely because A was constructed as a set of semigroup generators for G.

THEOREM 8.5* (J. W. Cannon). *Let X be a finite set of generators for the hyperbolic group G and let $A = X \cup X^{-1}$. Let L be the set of geodesic words of A^*, so L is the set of labels of geodesic edge-paths in the Cayley graph $\Gamma_{G,X}$. Then L is a regular sublanguage of A^*.*

A short and very readable proof of this result which is patterned after an argument due to Thurston can be found in [BGSS] Theorem 6.2.

8.5. EXAMPLE. Let $F = F(c,d)$ be the free group with free basis $\{c,d\}$. Let $A = \{c, C, d, D\}$ and let $\mu\colon A^* \to F$ be given by $c \mapsto c$, $C \mapsto c^{-1}$, $d \mapsto d$, and $D \mapsto d^{-1}$. The geodesic words in the alphabet A are just the reduced words, where a word is called reduced if it contains no subword of the form cC, Cc, dD, or Dd. A FSA with alphabet A whose language is the set of reduced words is shown below in Figure 7. In this example every state is an accept state.

8.6. If G is a hyperbolic group and X is a finite set of generators for G then the Cayley graph $\Gamma_{G,X}$ is δ-hyperbolic for some $\delta \geq 0$. If $A = X \cup X^{-1}$, we have already seen that if γ, γ' are geodesic words in A^* considered as paths beginning at 1 in $\Gamma_{G,X}$ and if the end points of these geodesics are at most a unit apart, then there exists $k > 0$ so that $d\big(\gamma(i), \gamma'(i)\big) \leq k$ for all $i \in \mathbb{N}$. In fact we can take $k = 2(\delta + 1)$ by Lemma 5.1. This property of geodesics is called the *k-fellow traveler property*.

These properties of hyperbolic groups, a regular language of normal forms for all group elements possessing the k-fellow traveler property for some $k > 0$, will be abstracted to define the notion of an automatic structure in the next section.

9. Automatic Groups

9.1. Definition. An automatic structure (A, L) on a finitely generated group G is

(1) a finite set A of semigroup generators, so the evaluation map $\mu: A^* \to G$ is surjective, and a regular language $L \subset A^*$ so that $\mu(L) = G$, such that
(2) there exists a number $k > 0$ so that if $p, p' \in L$ are such that $d(\mu(p), \mu(p')) \leq 1$ in the word metric d of the Cayley graph $\Gamma_{G,A}$, then p and p', thought of as labels of paths starting at the same vertex of $\Gamma_{G,A}$, satisfy the k-fellow traveler property.

We can think of L as a set normal forms for elements of G, and (1) says we have a regular language of normal forms for all group elements. By (2) if the paths p, $p' \in L$ begin at the same vertex and end at most a unit apart, then $d(p(i), p'(i)) \leq k$ for all $i \in \mathbb{N}$ (recall the convention 5.0 about extending the domain of edge-paths to $\mathbb{N}$). In particular, there may be many normal forms for the same group element, but all such are k-fellow travelers.

9.2. Examples.

9.2.1. The theorem of Cannon assures us that if A is a symmetric set of generators for the hyperbolic group G, then the set L of those words in A^* which are labels of geodesics in $\Gamma_{G,A}$ is an automatic structure on G.

9.2.2. Let $G = \mathbb{Z}^2$ with semigroup generators $c = (1, 0)$, $C = (-1, 0)$, $d = (0, 1)$, and $D = (0, -1)$, so $A = \{c, C, d, D\}$. Let $L = c^*d^* + c^*D^* + C^*d^* + C^*D^*$. Then (A, L) is an automatic structure for G. Those who know something about regular languages will recognize that L is given by a "regular expression", which defines a regular language. However it is easy enough to see that L is the language recognized by the FSA in Figure 8; Y is the set of all states in this machine.

In terms of the square lattice lattice in $\mathbb{R}^2$, the language L can be described as the language of geodesics doing all horizontal movement before doing any vertical movement. Note that each group element has a unique normal form in this language.

It remains to verify the k-fellow traveler property for suitable $k > 0$. We shall carry out the verification for $p, p' \in L$ representing points is the first quadrant; the other cases work the same way. Consider then $p = c^m d^n$ and $p' = c^{m+1} d^n$ where $m, n \geq 0$ thought of as paths from the origin. These paths agree for $t \leq m$, then diverge. But for $t = i > m$, $i \in \mathbb{N}$, $i \leq m + n$, we have $p(i) = c^m d^{i-m}$ and $p'(i) = c^{m+1} d^{i-m-1}$. We see that the distance in the word metric between these group elements is 2, so the paths are 2-fellow travelers. There is also the case $p = c^m d^n$ and $p' = c^m d^{n+1}$ to consider, but it is clear these paths are 1-fellow travelers. Thus $k = 2$ and the language L satisfy the 2-fellow traveler property.

9.3. In order to see that the property of having an automatic structure is independent of generators, one shows that an automatic structure (A, L) defined for one set of semigroup generators for G can be translated into an automatic structure for a second finite set A' of semigroup generators. The verifications are carried out in [**Wordproc**]. It is then permissible to define a finitely generated

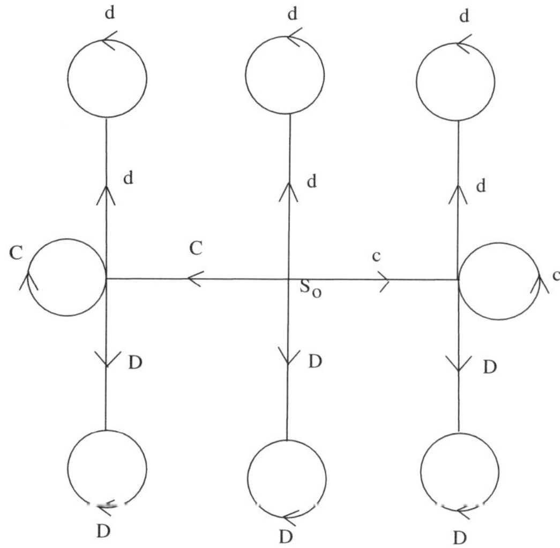

FIGURE 8. Automatic language for $\mathbb{Z}^2$.

group as being automatic if for one, and hence for every, finite set A of semigroup generators there is a regular language L of normal forms for all group elements satisfying the k-fellow traveler property for some $k > 0$.

9.4. Automatic groups are closed under the operations of taking finite index subgroups and finite extensions. The direct product and free product of two automatic groups is also automatic. Curiously, it is unknown whether a direct factor of an automatic group is automatic, although a free factor of one is always automatic. These facts are proved in [**Wordproc**] and in [**BGSS**]. Another curiosity is that in January 1989 we held a seminar at MSRI on automatic groups and I edited a set of problems from that seminar which is published in [**Gerst3**]; all those problems as originally posed are still open, although partial results have been obtained, and some of the problems have been solved in the setting of biautomatic groups, which I will not have time to treat here.

9.4.1. Example*. Consider the Euclidean Escher diagram Figure 1. The symmetry group G of this tesselation of $\mathbb{E}$ has a precompact fundamental domain, so it follows from the Bieberbach theorem[9] that G contains a subgroup of finite index isomorphic to $\mathbb{Z}^2$. It follows from 9.4 that G is automatic.

9.5. Automatic groups have been appearing sporadically in odd contexts. All lattices in the Lie groups $\mathrm{SO}(n, 1)$, all Coxeter groups, and Artin groups associated to the finite Coxeter groups are automatic. Perhaps the most striking result in the area, due to L. Mosher, is that all mapping class groups of surfaces of finite type (*i.e.* closed surfaces with a finite number of punctures) are automatic. However $\mathrm{Out}(F_3)$ is not automatic, as was shown by K. Vogtmann, basing her work on Thurston's argument that $\mathrm{Sl}_3(\mathbb{Z})$ is not automatic. It remains to be clarified where this notion fits into the general problem of understanding finitely presented groups.

[9]This states that if G is a subgroup of the group of isometries of Euclidean space $\mathbb{E}^n$ which acts properly discontinuously and cocompactly, then G contains a subgroup of translations which is of finite index in G and is isomorphic to $\mathbb{Z}^n$.

9.6. Automatic groups are finitely presented. In fact, the proof I gave in § 5 that hyperbolic groups are finitely presented only used the k-fellow traveler property, and hence applies without change to automatic groups. It follows from that argument that if (A, L) is an automatic structure for G with L satisfying the k-fellow traveler property, then $\mathcal{P} = \langle A \mid \mathcal{R} \rangle$ is a finite presentation for G, where $\mathcal{R}$ is the set of labels of all edge-loops in the Cayley graph $\Gamma_{G,A}$ of length at most $2k + 2$.

It might appear that the same argument would prove the quadratic isoperimetric inequality, but there is a sticking point. In the case of hyperbolic groups, the coning argument in Theorem 5.3 used geodesics, and one always has a bound on the lengths of geodesics, and hence a bound for the number of relators Q_{ij} there. However if we try to cone using the normal forms p in a regular language, we quickly find that there is no bound in general on the length of p in terms of $d\big(1, \mu(p)\big)$. I want to spend the remaining time showing how to fix this problem and proving

9.7. THEOREM. *An automatic group satisfies the quadratic isoperimetric inequality.*

The proof makes use of two lemmas. We say that (A, L') is an automatic substructure of the automatic structure (A, L) for the group G if L' is a regular sublanguage contained in L which is mapped onto G. If L has the k-fellow traveler property, it follows that L' also has the k-fellow traveler property with the same k.

9.8. LEMMA. *Each automatic structure (A, L) for the group G contains an automatic substructure (A, L'), whose language L' is mapped bijectively by the evaluation mapping onto G.*

One orders the alphabet A and takes the lexicographically least representative for each group element in L. This defines L' and it is clearly mapped bijectively onto G by the evaluation mapping. For the proof that L' is regular, see [**Wordproc**] 2.5 or [**BGSS**] Prop. 1.3 page 255.

9.9. LEMMA. *If (A, L) is an automatic structure for the group G such that the evaluation mapping μ maps L bijectively onto G, then there exists $M > 0$ so that for all pairs $w, w' \in L$ with $d\big(\mu(w), \mu(w')\big) = 1$ one has $\big|\ell(w) - \ell(w')\big| \leq M$.*

PROOF. Assume the conclusion is false, so that one can find pairs of words $w, w' \in L$ which evaluate in G a unit apart with w' arbitrarily longer than w. Let $\mathcal{A}$ be a FSA with alphabet A which recognizes the language L, and let L satisfy the k-fellow traveler property. If $\ell(w) = n$, then for all $i > 0$ one has $d\big(w(n), w'(n+i)\big) \leq k$. But the ball of radius k about 1 in $\Gamma_{G,A}$ is finite and there are only a finite number of states in the FSA $\mathcal{A}$, so if $\ell(w') \gg \ell(w)$, then there must exist integers $i < j$ with $0 < i < j < \ell(w') - \ell(w)$ so that $w'(j)$ and $w'(i)$ represent the same element of G *and* such that the word w' is in the same state of the machine $\mathcal{A}$ at time i as at time j. But then the string w' makes a loop in the machine between times i and j and the label of this loop is the identity element of G. However this means that you can go around this loop any number of times and still get to an accept state of $\mathcal{A}$, thereby generating the same element $\mu(w')$ of G infinitely often. In other words, L is not mapped bijectively onto G by μ, a contradiction.

It follows that the difference $\big|\ell(w) - \ell(w')\big|$ is bounded if one considers all pairs $w, w' \in L$ which evaluate a unit apart in G, and the Lemma is established.

We can now give the proof of Theorem 9.7. By Lemma 9.8 there is an automatic structure (A, L) for G where the evaluation mapping maps L bijectively onto G. By Lemma 9.9 elements of L which evaluate to group elements a unit apart differ in length by no more that some fixed constant M. Assume that L satisfies the k-fellow traveler property.

From 9.6 we know that the set of labels of all loops in $\Gamma_{G,A}$ of length at most $2k+2$ is a defining set of relations among the generators A. Suppose now that w is an edge-loop in $\Gamma_{G,A}$ with $w(0) = 1$. We cone from the identity element, in this case not by geodesics, but rather we choose for each $0 \leq i \leq \ell(w)$ the unique element $p_i \in L$ which evaluates to $w(i)$ in G. The same picture, Figure 5, applies here and may help to visualize the argument. Using Lemma 9.9 and induction we see that $\ell(p_i) \leq Mi$ for each i. It follows that the strip between p_i, p_{i+1} and the edge of w has area at most Mi. Thus we can fill w using at most $\sum_{i=0}^{\ell(w)-1} Mi \leq C\ell(w)^2$ defining relators, where C is a constant independent of w. This establishes the quadratic isoperimetric inequality.

References

[Alon] J. Alonso, *Inégalités isopérimétriques et quasi-isométries*, C. R. Acad. Sci. Paris, Sér. I **311** (1991), 761–764.

[Bkly] J. Alonso, T. Brady, D. Cooper, V. Ferlini, M. Lustig, M. Mihalik, M. Shapiro, and H. Short, *Notes on word hyperbolic groups*, Group Theory from a Geometrical Viewpoint (E. Ghys, A. Haefliger, and A. Verjovsky, eds.), World Scientific, Singapore, 1991, pp. 3–63.

[BGSS] G. Baumslag, S. M. Gersten, M. Shapiro, and H. Short, *Automatic groups and amalgams*, J. Pure Appl. Algebra **76** (1991), 229–316.

[BMS] G. Baumslag, C. F. Miller III, and H. Short, *Isoperimetric inequalities and the homology of groups*, Invent. Math. **113** (1993), 531–560.

[Beard] A. F. Beardon, *The eometry of discrete groups*, Graduate Texts in Mathematics, vol. 91, Springer-Verlag, New York-Berlin, 1983.

[Bndry] M. Bestvina, *Boundary of hyperbolic groups* (notes on seminar lectures taken by J. Burillo: http://www.math.utah.edu/~gersten/MaxDehnSeminar/burillo.dvi).

[Brady] N. Brady, *Branched coverings of cubical complexes and subgroups of hyperbolic groups*, Univ. of Utah (preprint).

[Wordproc] J. W. Cannon, D. B. A. Epstein, D. F. Holt, S. V. F. Levy, M. S. Paterson, and W. P. Thurston, *Word processing in groups*, Jones and Bartlett, Boston, 1992.

[Stras] M. Coornaert, T. Delzant, and A. Papadopoulos, *Notes sur les groupes hyperboliques de Gromov*, Lecture Notes in Math., vol. 1441, Springer-Verlag, 1990.

[Gerst] S. M. Gersten, *Dehn functions and ℓ_1-norms of finite presentations*, Algorithms and Classification in Combinatorial Group Theory (G. Baumslag and C. F. Miller III, eds.), Math. Sci. Res. Inst. Publ., vol. 23, Springer-Verlag, New York, 1992, pp. 195–224.

[Gerst2] ———, *Subgroups of word hyperbolic groups in dimension* 2, J. London Math. Soc. (2) **54** (1996), no. 2, 261–283 (http://www.math.utah.edu/~gersten).

[Gerst3] ———, *Problems on automatic groups*, Algorithms and Classification in Combinatorial Group Theory (G. Baumslag and C. F. Miller III, eds.), Math. Sci. Res. Inst. Publ., vol. 23, Springer-Verlag, 1992, pp. 225–232.

[Swiss] E. Ghys and P. de la Harpe (eds.), *Sur les groupes hyperboliques d'après Mikhael Gromov*, Progr. Math., vol. 83, Birkhäuser, 1990.

[Gromov] M. Gromov, *Hyperbolic groups*, Essays in Group Theory (S. M. Gersten, ed.), Math. Sci. Res. Inst. Publ., vol. 8, Springer-Verlag, 1987, pp. 75–263.

[L-S] R. C. Lyndon and P. E. Schupp, *Combinatorial group theory*, Springer-Verlag, 1977.

[Olshan] A. Yu. Ol'shanskii, *Hyperbolicity of groups with subquadratic isoperimetric inequality*, Internat. J. Algebra Comput. **1** (1991), 281–290.

[Papa] P. Papasoglu, *Strongly geodesically automatic groups are hyperbolic*, Inv. Math. **121** (1995), 323–334.

[Rips] E. Rips, *Subgroups of small cancellation groups*, Bull. London Math. Soc. **14** (1982), 45–47.

Mathematics Department, University of Utah, Salt Lake City, UT 84112, USA
E-mail address: `gersten@math.utah.edu`

Centre de Recherches Mathématiques
CRM Proceedings and Lecture Notes
Volume **17**, 1999

Description of Fully Residually Free Groups and Irreducible Affine Varieties Over a Free Group

O. Kharlampovich and A. Myasnikov

The object of this talk is to describe irreducible varieties over free groups and to characterize finitely generated fully residually free groups. We prove that any variety over a free group F can be defined by a finite number of systems of equations $S = 1$ in the triangular form where quadratic words play the role of leading terms. Algebraically, irreducible varieties are exactly the varieties whose coordinate groups are discriminated by F. The crucial point of the classification of fully residually free groups is to prove that the coordinate groups of irreducible varieties are embeddable into Lyndon's free $\mathbf{Z}[x]$-group $F^{\mathbf{Z}[x]}$. Since every finitely generated fully residually free group is a free factor of the coordinate group of an irreducible variety, and the group $F^{\mathbf{Z}[x]}$ is fully residually F, we obtain a characterization of finitely generated fully residually free groups as subgroups of $F^{\mathbf{Z}[x]}$.

The group $F^{\mathbf{Z}[x]}$ and its subgroups have been intensively studied during the last several years. In particular, every finitely generated subgroup of $F^{\mathbf{Z}[x]}$ (hence, every finitely generated fully residually free group) can be obtained from free abelian groups of finite rank by finitely many free products with amalgamation and HNN-extensions of very special type in which amalgamated and associated subgroups are free abelian of finite rank. In particular, it implies that every finitely generated fully residually free group is finitely presented.

An $\exists$-free group is a group G such that the class of $\exists$-formulas, true in G, is the same as the class of $\exists$-formulas, true in a nonabelian free group. A finitely generated group is $\exists$-free if and only if it is fully residually free [**18**]. Our result gives an algebraic description of $\exists$-free groups.

The result also implies that $F^{\mathbf{Z}[x]}$ is algebraically closed with respect to the universal theory of a free group.

There are three key parts to this work: algebraic geometry over free groups, the theory of free exponential groups, and Makanin-Razborov's machinery to deal with equations over free groups [**12, 15, 14**].

The algebraic geometry approach has been proved to be very useful when we deal with equations over groups. It provides necessary topological means and a method to transcribe geometric notions into pure group-theoretic language. Following Baumslag, Myasnikov, Remeslennikov [**1**] we use the standard algebraic

---

1991 *Mathematics Subject Classification.* Primary: 20E05; Secondary: 20F10.
This is the final form of the paper.

geometry notions such as variety, Zariski topology, irreducibility of varieties, radicals and coordinate groups. Some of the ideas of the algebraic geometry approach go back to R. Lyndon [**10**], E. Rips, J. Stallings [**20**], V. Guba [**8**], B. Plotkin.

The theory of exponential groups (i.e. groups admitting exponents in some ring A) starts with results of P. Hall, A. Malcev, G. Baumslag and R. Lyndon. It provides a technique to deal with noncommutative modules over the ring A. R. Lyndon gave an axiomatic description of the notion of exponential group. He described and studied the group $F^{\mathbf{Z}[x]}$ (free exponential group over the ring of integral polynomials $\mathbf{Z}[x]$). He showed the crucial importance of this group in the study of equations over free groups. A modern treatment of exponential groups was given by Myasnikov and Remeslennikov in [**13**]. In particular they showed that $F^{\mathbf{Z}[x]}$ can be described using HNN-extensions of very special type, namely extensions of centralizers. Basically, to obtain $F^{\mathbf{Z}[x]}$ from F one needs just extend all centralizers of F up to free $\mathbf{Z}[x]$-modules of rank 1. Namely, $F^{\mathbf{Z}[x]}$ is the union of groups:

$$F < G_1 < G_2 < \cdots,$$

where $G_{i+1} = \langle G_i, t \mid C(u_i)^t = C(u_i)\rangle$, where $C(u_i)$ is some proper centralizer in C_i.

The groups $F^{\mathbf{Z}[x]}$ happen to be fully residually F [**9**]. Basically there exists only one known method of proving that a group is fully residually free. This method is due to G. Baumslag who showed that the surface groups (except the non-orientable case of genus 1,2,3) are embeddable into an extension of a centralizer of a free group, and hence they are fully residually free. B. Baumslag gave other examples of fully residually free groups using the same method.

In 1992 Myasnikov and Remeslennikov studying ultrapowers of free groups came to the following conjecture: every finitely generated fully residually free group is a subgroup of Lyndon's group $F^{\mathbf{Z}[x]}$.

In the paper [**5**] the conjecture was proved for 3-generated fully residually free groups, moreover it was proved that there are just three types of 3-generated fully residually free groups: free groups, free abelian groups, and extensions of centralizers $\langle x, y, t \mid u^t = v\rangle$, where u is an arbitrary element in $F(x, y)$ and is not a proper power.

First we concentrate on quadratic equations over a free group F. Quadratic equations have been widely studied, see for example [**7, 6, 4, 3, 11**].

We shall prove that the coordinate group $F_{R(S)}$ of the quadratic equation $S = 1$ is embeddable into $F^{\mathbf{Z}[x]}$. It implies that the variety $V(s)$ is irreducible in Zariski topology over F^n. Moreover, we completely describe the radical of the quadratic system $S = 1$. It turns out that the radical $R(S)$ coincides (with a few exceptions) with the normal closure of S in the group $F * F(X)$. In particular, it implies the Nullstellensatz for the system $S = 1$. The group formulation of Nullstellensatz was given by E. Rips, who actually announced in a seminar in New York in the fall of 1995, the joint result with Z. Sela about the Nullstellensatz for quadratic equations over a free group.

We shall give an algorithm to represent a solution set of arbitrary system of equations over F as a union of finite number of irreducible components in Zariski topology on F^n. The solution set for every system is contained in the solution set of a finite number of systems in triangular form with quadratic words as leading terms. The possibility of such a decomposition for a solution set was conjectured by Razborov in [**16**] and also by Rips.

We shall give a description of systems of equations determining irreducible components. We shall use methods developed in [**12**] and [**15**], it is possible to find some of them in [**14**]. We are thankful to E. Rips for attracting our attention to these methods. We will follow the terminology given in [**1**].

Let G be a group, $F = F(X)$ the free group with a basis $X = \{x_1, x_2, \dots\}$, $G[X] = G * F$ free product of G and F.

An element s from $G[X]$ is called *an equation over group* G. We write this as $s = 1$. As an element of the free product, s can be written as a product of some elements $x_1, \dots, x_n$ from $X \cup X^{-1}$ (which are called *variables*) and elements $g_1, \dots, g_m$ from G (*constants*). We will write, sometimes, $s(x_1, \dots, x_n, g_1, \dots, g_m) = 1$ or, simply, $s(x, g) = 1$. *A system of equations over group* G is an arbitrary set of equations $\mathcal{S} = \{s_i = 1 \mid i \in I\}$ (Notation: $\mathcal{S} = 1$). *A solution* of a system $\mathcal{S}(x_1, \dots, x_n, g_1, \dots, g_m) = 1$ over a group G is a tuple of elements $a_1, \dots, a_n \in G$ such that after replacement of each x_i by a_i in every equation $s(x, g) = 1$ one gets a trivial element in a group G. On the other hand, a solution of the system $\mathcal{S} = 1$ over G can be described as a retract $\phi\colon G[X] \to G$ such that $\phi(\mathcal{S}) = 1$. These definitions are equivalent. By $V(\mathcal{S})$ we denote the set of all solutions in G of the system $\mathcal{S} = 1$.

Let S be a subset of $G[X]$. Then $V(S)$ is called an *algebraic subset* or an (affine) variety in G^n. Two systems $S = 1$ and $T = 1$ are equivalent over G if $V(S) = V(T)$. For any $S \subseteq G[X]$ we have $V(S) = V\big(\mathrm{ncl}(S)\big)$, where $\mathrm{ncl}(W)$ is the normal closure of S in $G[X]$.

A group G is called CSA-group if every maximal abelian subgroup M of G is malnormal, i.e. $M^g \cap M = 1$ for any $g \notin M$.

It was shown in [**1**] that for a nonabelian CSA-group G all algebraic sets in G^n define a topology on G^n in which they are exactly the closed sets. (We have $V(S) = V\big(\mathrm{ncl}(S)\big)$, $V(\bigcup S_i) = \bigcap V(S_i)$, $V(S_i) \cup V(S_2) = V([u^{a^\alpha}, v^{b^\beta}],\ u \in S_1, v \in S_2, a, b \in G, [a, b] \neq 1, \alpha, \beta \in \{1, -1\})$, $V(1 = 1) = G^n$ and $V(G[X]) = \emptyset$.) The topology defined by algebraic sets as closed subsets is said to be a Zariski topology.

Below G is always a nonabelian group.

Definition 1. Let $Y \subseteq G^n$. Define a set

$$I(Y) = \big\{s \in G[X] \mid s(g_1, \dots, g_n) = 1 \forall (g_1, \dots, g_n) \in Y\big\}.$$

The set $I(Y)$ has a nice description in terms of homomorphisms. Any tuple $g = (g_1, \dots, g_n) \in Y$ defines a homomorphism $f_g\colon G[X] \to G$ by the condition $x_i \to g_i$. Then

$$I(Y) = \bigcap_{g \in Y} \ker(f_g).$$

Let us recall that a subgroup N is *an isolated subgroup* in a group H if for any $x \in H$ and any nonzero integer n inclusion $x^n \in N$ implies that $x \in N$. For any set $S \subset H$ the intersection of all *normal isolated* subgroups containing S is denoted by $\sqrt{S}$.

Lemma 1 ([**1**]). (1) *$I(Y)$ is a normal subgroup of $G[X]$;*

(2) *If G is a torsion-free group then $I(Y)$ is an isolated normal subgroup of $G[X]$, in particular $I\big(V(S)\big)$ contains $\sqrt{S}$.*

DEFINITION 2. Let $V(S)$ be a variety defined by $S \subset G[X]$. Then $I(V(S))$ is called the radical of the system $S = 1$ and is denoted by $\mathrm{Rad}(S)$. The quotient group $G_{R(S)} = G[X]/R(S)$ is called the affine coordinate group of the variety $V(S)$.

Systems $S = 1$ and $T = 1$ define the same variety over G iff $\mathrm{Rad}(S) = \mathrm{Rad}(T)$.

Let $S = 1$ be a system of equations over a torsion-free group G. Then the quotient group of $G[X]$ by the $\sqrt{S}$ is denoted by $G_{\sqrt{S}}$.

A system $S = 1$ over G is called consistent if there is a homomorphism $\pi\colon G[X] \to H \geq G$ which is a retract on G such that $S \in \ker(\pi)$. Otherwise it is inconsistent over G. If a system $S = 1$ over G is consistent then the canonical homomorphism $G \to G_S$ is monic. An inconsistent system $S = 1$ defines the empty variety over G. Therefore, for non-empty varieties $V(S)$ we will assume that G is a subgroup of $G_{R(S)}$.

Let G be a torsion-free group, $V(S)$ is an algebraic set in G^n defined by a system $S = 1$. Then by the lemma $I(V(S))$ contains $\sqrt{S}$.

DEFINITION 3. A system of equations $S = 1$ over group G satisfies the Nullstellensatz if

$$I(V(S)) = \sqrt{S}.$$

DEFINITION 4. Let H be a group and $\mathcal{G}$ be a family of groups.

(1) A homomorphism of groups $\psi\colon H \to G$ separates a nontrivial element $h \in H$ if $\psi(h) \neq 1$;
(2) A family of homomorphisms $\Psi = \{\psi\colon H \to G \mid G \in \mathcal{G}\}$ is called a separating (discriminating) family of homomorphisms if any nontrivial $h \in H$ (any finite number of nontrivial elements $h_1, \dots, h_n \in H$) can be separated by some $\psi \in \Psi$. In this case H is called a residually $\mathcal{G}$ group (ω-residually $\mathcal{G}$ group or fully residually G group).

In the case when $\mathcal{G} = G \leq H$, and homomorphism $\phi\colon H \to G$ is identical on G, ϕ is called a G-homomorphism. If there is a separating (discriminating) family of G-homomorphisms $H \to G$, we say that H is separated (discriminated) by G-homomorphisms.

LEMMA 2 ([**1**]). *A system of equations $S = 1$ over a torsion-free group G satisfies the Nullstellensatz in G if and only if $G_{\sqrt{S}}$ is separated by G-homomorphisms.*

Denote by G_S the factor group $G[X]/\,\mathrm{ncl}(S)$.

LEMMA 3. *If G is torsion free and G_S is separated by G-homomorphisms, then* $\mathrm{ncl}(S) = \sqrt{S} = \mathrm{Rad}(S)$.

The proof is straightforward.

A group G is called *Equationally Noetherian* (EN) if for every system S of equations over G there is a finite subsystem S_0 such that $V(S) = V(S_0)$.

A closed set in Zariski topology over an EN group is called *irreducible* if it cannot be represented as a union of two proper closed subsets.

LEMMA 4. *Let G be* ENCSA*-group. Then $V(S)$ is irreducible if and only in $G_{R(S)}$ is discriminated by G-homomorphisms.*

PROOF. Suppose $V(S)$ is not irreducible, $V(S) = \bigcup_{i=1}^{n} V(S_i)$. Then there exist $s_i \in S_i \setminus \{\mathrm{Rad}(S), \mathrm{Rad}(S_j), j \neq i\}$. The set $s_i, i = 1, \dots, n$ cannot be separated. $\square$

Suppose now $s_1, \dots, s_m$ are elements such that for any retract $f\colon G_{R(S)} \to G$ there exist i such that $f(s_i) = 1$, then $V(S) = \bigcup_{i=1}^{n} V(S \cup s_i)$.

DEFINITION 5. Let K be a group, $C(u)$ the centralizer of an element $u \in K$. Suppose $C(u)$ is abelian. Then the following group is called a free extension of a centralizer in K:

$$K(u,t) = \langle K,\ t \mid [C(u), t] = 1 \rangle.$$

Note, that $K(u,t)$ can be obtained from K by an HNN-extension with respect to the identity isomorphism $C(u) \to C(u)$:

$$G(u,t) = \langle G, t \mid t^{-1}at = a,\ a \in C(u) \rangle.$$

We introduce the following notation. Let

$$G = G_0 \leq G_0(u_1, t_1) = G_1 \leq \cdots \leq G_n(u_n, t_n) = G_{n+1}$$

be a finite sequence of extensions of centralizers of elements $u_i \in G_i$. Then we denote the resulting group G_{n+1} by $G(U,T)$, where $U = \{u_1, \dots, u_n\}$, $T = \{t_1, \dots, t_n\}$.

Let A be an arbitrary associative ring with identity and G a group. Fix an action of the ring A on G, i.e. a map $G \times A \to G$. The result of the action of $\alpha \in A$ on $g \in G$ is written as g^α. Consider the following axioms:

(1) $g^1 = g$, $g^0 = 1$, $1^\alpha = 1$;
(2) $g^{\alpha+\beta} = g^\alpha \cdot g^\beta$, $g^{\alpha\beta} = (g^\alpha)^\beta$;
(3) $(h^{-1}gh)^\alpha = h^{-1}g^\alpha h$;
(4) $[g,h] = 1 \implies (gh)^\alpha = g^\alpha h^\alpha$.

DEFINITION 6. Groups with A-actions satisfying axioms (1)–(4) are called *A-groups.*

In particular, an arbitrary group G is a $\mathbf{Z}$-group. We now recall the definition of A-completion.

DEFINITION 7. Let G be a group. Then an A-group G^A together with a homomorphism $G \to G^A$ is called a *tensor* A-completion of the group G if G^A satisfies the following universal property: for any A-group H and a homomorphism $\varphi\colon G \to H$ there exists a unique A-homomorphism $\psi\colon G^A \to H$ (a homomorphism that commutes with the action of A) such that the following diagram commutes:

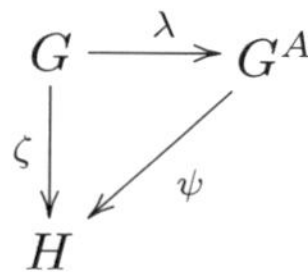

By $\mathbf{Z}[x]$ we denote as usual the ring of polynomials of one variable with integer coefficients.

LEMMA 5 ([**13**]). *Every group obtained from a* CSA*-group G by a sequence of free extensions of centralizers is embeddable into $G^{\mathbf{Z}[x]}$.*

1. Quadratic Equations Over a Free Group

NOTATION. Let $S \in G[X]$, then the set of all variables which occur in S is denoted by $\text{var}(S)$.

Two systems $S = 1$ and $T = 1$ over G are termed to be *disjoint* if $\text{var}(S) \cap \text{var}(T) = \emptyset$. A system $S = 1$ is *splittable* if it is a union of two nonempty disjoint subsystems: $S = S_1 \cup S_2$, $\text{var}(S) \cap \text{var}(T) = \emptyset$.

DEFINITION 8. A set $S \subset G[X]$ is called quadratic, if every variable from $\text{var}(S)$ occurs in S not more then twice. The set S is strictly quadratic if every letter from $\text{var}(S)$ occurs in S exactly twice.

A system $S = 1$ over G is *quadratic* (*strictly quadratic*), if the corresponding set S is quadratic (strictly quadratic).

THEOREM 1. *Let G be a fully residually free group and let $S = 1$ be a quadratic equation over G, then $G_{R[S]}$ is G-embeddable into $G(U,T)$ for some sequence of extensions of centralizers, and hence into $G^{\mathbf{Z}[x]}$.*

DEFINITION 9. A system $\bigcup_{i=1}^{m} S_i(\bar{c}, \bar{x}_1, \ldots, \bar{x}_i) = 1$ is said to be triangular quasi-quadratic if for every i the equation $S_i(\bar{c}, x_i, \ldots, x_n) = 1$ is quadratic in variables $\bar{x}_i$

Such a system is said to be nondegenerate if for each i the equation $S_i = 1$ over $G[\bar{x}_{i+1}, \ldots, \bar{x}_m]/R(\bigcup_{j=1-1}^{m} S_j)$ (with $\bar{x}_i$ considered as variables and $\bar{c}, \bar{x}_{i+1}, \ldots, \bar{x}_m$ as coefficients) has a solution.

The following result is a corollary of Theorem 1.

THEOREM 2. *If S is a nondegenerate triangular quasi-quadratic system over a fully residually free group G, then $G_{R(S)}$ is a subgroup of $G(U,T) \leq G^{\mathbf{Z}[x]}$.*

COROLLARY 1. *Algebraic sets corresponding to triangular quasi-quadratic systems of equations are irreducible sets in Zariski topology on G^n.*

Every quadratic equation over G can be transformed into a standard equation:

DEFINITION 10. A standard quadratic equation over group G is an equation of the one of the following forms:

$$\prod_{i=1}^{n}[x_i, y_i] = 1, \quad n > 0; \tag{1}$$

$$\prod_{i=1}^{n}[x_i, y_i] \prod_{i=1}^{m} z_i^{-1} c_i z_i = d, \quad n, m \geq 0, m + n \geq 1; \tag{2}$$

$$\prod_{i=1}^{n} x_i^2 = 1, \quad n > 0; \tag{3}$$

$$\prod_{i=1}^{n} x_i^2 \prod_{i=1}^{m} z_i^{-1} c_i z_i = d, \quad n, m \geq 0, n + m \geq 1; \tag{4}$$

where d, c_i are nontrivial elements from G.

LEMMA 6. *Let S be a strictly quadratic word over G. Then there is a G-automorphism $f \in \text{Aut}_G(G[X])$ such that S^f is a standard quadratic word over G.*

PROOF. See [**3**]. □

In this case we say that S is equivalent to S^f over $G[X]$.

DEFINITION 11. Strictly quadratic words of the type

$$[x,y],\ x^2,\ z^{-1}cz$$

where $c \in G$, are called *atomic quadratic words* or simply *atoms*.

DEFINITION 12. Any standard quadratic equation $S = 1$ over G can be written (see above) as a product of atoms r_i:

$$r_1\, r_2 \dots r_k = g.$$

The number k is called the atomic rank of S. (Notation: $k = r(S)$).

DEFINITION 13. A solution ϕ of a quadratic equation of atomic rank $k \geq 2$ is called commutative if $[r_i^\phi, r_{i+1}^\phi] = 1$ for all $i = 1, \dots, k-1$, otherwise it is called noncommutative.

THEOREM 3. *Let G be a fully residually free group and let $S = 1$ be a standard quadratic equation over G. Then*

(1) *if the atomic rank of $S = 1$ is greater than 1 and $S = 1$ has a noncommutative solution, or $S = [x,y]d^{-1} = 1$, or $S = [x,y][z,t] = 1$, then* $\mathrm{Rad}(S) = \mathrm{ncl}(S)$,
(2) *if all solutions of $S = 1$ are commutative and $S \neq [x,y][z,t] = 1$, or $S = c^z d^{-1} = 1$, then up to some linear transformation of variables* $\mathrm{Rad}(S) = \mathrm{ncl}\{[u_i, b_i] = 1 \mid i = 1, \dots, k\}$, *where G_i's are constants from G,*
(3) *in the following cases S always has a noncommutative solution, hence* $\mathrm{Rad}(S) = \mathrm{ncl}(S)$:
    (a) *S has type* (1), $n > 2$,
    (b) *S has type* (2), $n > 0, n + m > 1$ *(in case $n = 1, m = 0$ the notion of noncommutative solution is not defined, but* $\mathrm{Rad}(S) = \mathrm{ncl}(S)$*),*
    (c) *S has type* (3), $n > 3$,
    (d) *S has type* (4), $n > 2$.

LEMMA 7. *Let S be a quadratic system over G. Then there exists a system $S' = \{s_1, \dots, s_n\}$ of strictly quadratic pairwise disjoint equations $s_1, \dots, s_n$ such that*

$$G_S \simeq_G G_{S'}.$$

PROOF. Suppose a letter $x \in \mathrm{var}(S)$ occurs in some equation $s \in S$ and exactly once. Then we can rewrite the equation $s = 1$ in the form $x = s'$. Hence the group G_S can be Tietze transformed to the G-isomorphic group G_{S^*}, where S^* is obtained from S by deleting the equation $s = 1$ and replacing all other occurrences of x by s'. After a finitely many transformations of this type we arrive to a system S' such that G_S is G-isomorphic to $G_{S'}$ and every variable from $\mathrm{var}(S')$ occurs in one and only one equation from S' and exactly twice. It follows that S' is a system of pairwise disjoint strictly quadratic equations over G. □

The following result can be deduced from the proof of Lemma 6 in [**3**] and Lemma 7

LEMMA 8. *Let S be a quadratic system over G. Then there exists a system $S' = \{s_1, \dots, s_n\}$ of standard quadratic pairwise disjoint equations $s_1, \dots, s_n$ and a free group F_1 such that*

$$G_S \simeq_G G_{S'} * F_1.$$

A system $S' = 1$ from the lemma above splits over G in such a way that

$$G_S \simeq_G \big(\dots(G_{s_1})_{s_2}\dots\big)_{s_n} * F_1.$$

Hence we have

COROLLARY 2. *Let $S = 1$ be a quadratic system over G. Then*

$$G_S \simeq_G \big(\dots(G_{s_1})_{s_2}\dots\big)_{s_n} * F_1,$$

where s_i is a standard quadratic equation over G and F_1 is some finitely generated free group.

2. Irreducible Varieties

THEOREM 4. *For any finite system $S(\bar{x}) = 1$ over a free group F, one can find effectively a finite family of nondegenerate triangular quasi-quadratic systems $U_1, \dots, U_k$ and word mappings $p_i\colon V_F(U_i) \to V_F(S)$ $(i = 1, \dots, k)$ such that for every $b \in V_F(S)$ there exists i and $c \in V_F(U_i)$ for which $b = p(c)$, i.e.*

$$V_F(S) = p_1\big(V_F(U_1)\big) \cup \dots \cup p_k\big(V_F(U_k)\big)$$

and all sets $p_i\big(V_F(U_i)\big)$ are irreducible, moreover, every irreducible component of $V_F(S)$ can be obtained as a closure of some $p_i\big(V_F(U_i)\big)$ in Zariski topology.

THEOREM 5. *For a system $S = 1$ over a free group $V(S)$ is irreducible if and only if $F_{R(S)} \subseteq F_{R(S_1)}$ for a nondegenerate triangular quasi-quadratic system S_1.*

Notice that in [**2**] it was shown that $F^{\mathbf{Z}[x]}$ is fully residually free.
Theorem 5 implies

THEOREM 6. *A finitely generated group is fully residually free if and only if it is isomorphic to a subgroup of $F^{\mathbf{Z}[x]}$.*

PROOF. Consider a finitely generated fully residually free group G given by generators $x_1, \dots, x_n$ and relations $s_j(x_1, \dots, x_n), j \in J$. Consider $S = \{s_j(x_1, \dots, x_n), j \in J\}$ as a system of equations over F. Then $G = F_S$ and $\mathrm{ncl}(S) = \mathrm{Rad}(S)$, as F_S is fully residually free. Hence $G = F_{R(S)}$ is embeddable into $F^{\mathbf{Z}[x]}$. □

Now we can describe the algebraic structure of finitely generated subgroups of $F^{\mathbf{Z}[x]}$ in terms of free constructions.

Let $H = \langle G, t \mid A^t = B\rangle$ be an HNN-extension of G with associated subgroups A and B. It is called a *separated* HNN-*extension* if for any $g \in G$, $A^g \cap B = 1$.

COROLLARY 3. *Every finitely generated residually free group G is a subgroup of a direct product of finitely many fully residually free groups; hence G is embeddable into $F^{\mathbf{Z}[x]} \times \dots \times F^{\mathbf{Z}[x]}$.*

THEOREM 7. *Let V be an irreducible variety over F. Then there exists a finite system of equations $S = 1$ over F which defines the variety V and satisfies the Nullstellensatz.*

THEOREM 8 (joint with V. Remeslennikov). *Let a group G be obtained from a free group F by a series of finitely many free extensions of centralizers. Then every finitely generated subgroup H of G is obtained from free abelian groups of finite rank by finitely many operations of the following type:*

(1) *free products;*

(2) *amalgamated products with abelian amalgamated subgroups at least one of which is maximal abelian;*
(3) *free extensions of centralizers;*
(4) *separated* HNN*-extensions with abelian associated subgroups at least one of which is maximal abelian.*

COROLLARY 4. *Every finitely generated fully residually free group is finitely presented.*

Corollary 4 was also announced by Z. Sela.

COROLLARY 5. *All finitely generated subgroups of $F^{\mathbf{Z}[x]}$ in which all proper centralizers are cyclic, are hyperbolic.*

COROLLARY 6. *Every finitely generated group H which is $\forall\,\exists$-equivalent to a nonabelian free group is torsion-free hyperbolic, moreover H can be obtained from infinite cyclic groups by finitely many operations of the following type:*
(1) *free products;*
(2) *amalgamated products with infinite cyclic amalgamated subgroups at least one of which is maximal abelian;*
(3) *separated* HNN*-extensions with infinite cyclic associated subgroups at least one of which is maximal abelian.*

In [**19**] Remeslennikov proved that every finitely generated fully residually free group acts freely on some $\mathbf{Z}^n$-tree with some order for a suitable natural number n. In [**17**] he asks (Question A) if such group acts freely on some $\mathbf{Z}^n$-tree with lexicographic order. Corollary 7 gives a positive answer to his question.

COROLLARY 7. *Every finitely generated fully residually free group acts freely on some $\mathbf{Z}^n$-tree, where $\mathbf{Z}^n$ is a direct sum of n copies of $\mathbf{Z}$ with lexicographic order.*

Let $U = \{u_1, \dots, u_n\}$ be a set of parametric words, i.e. a subset of $F^{\mathbf{Z}^k}$. By the definition we have fixed some pure cyclic subgroup $\mathbf{Z}$ in $\mathbf{Z}^k$ in such a way that the action of this subgroup $\mathbf{Z}$ coincides with the integer powers in F. Because $\mathbf{Z}$ is pure in $\mathbf{Z}^k$ we have $\mathbf{Z}^k = \mathbf{Z} \oplus B$, where B is a free abelian group with a free base $t_1, \dots, t_n$. These generators t_i-s are called parameters in $F^{\mathbf{Z}^k}$. Any homomorphism $\xi\colon B \to \mathbf{Z}$ gives rise to a F-homomorphism $\xi^\star\colon F^{\mathbf{Z}^k} \to F$. In this case we say that the image $U^{\xi^\star}$ is obtained from U by specializing parameters by ξ. Let

$$U^\star = \bigcup \{U^{\xi^\star} \mid \xi \in \mathrm{Hom}(\mathbf{Z}^k, \mathbf{Z})\}$$

be the union of all specializations of the set U.

We can slightly generalize the construction of a specialization. Instead of a set U we can consider a set of tuples of words from $F^{\mathbf{Z}^k}$ and specialize them coordinatewise. So we will get the set $U^\star$ of tuples of elements from F.

THEOREM 9. *Let $S(X) = 1$ be a system of equations over a free group F. Then there exists a finite set of n-tuples of parametric words $U = (u_1, \dots, u_n) \in (F^{\mathbf{Z}^k})^n$ such that the set of all their specializations $U^\star$ is a dense subset of the variety $V_F(S)$ in Zariski topology.*

COROLLARY 8. *Any system $S = 1$ over a free group F has a dense subset which can be parametrized by a finitely many parametric words.*

References

1. G. Baumslag, A. Myasnikov, and V. Remeslennikov, *Algebraic geometry over groups*, 1996 (to appear).
2. ______, *Residually hyperbolic groups and approximation theorems for extensions of centralizers*, 1996 (to appear).
3. L. P. Comerford and C. C. Edmunds, *Quadratic equations over free groups and free products*, J. Algebra **68** (1981), 276–297.
4. ______, *Solutions of equations in free groups*, Walter de Gruyter, Berlin, New York, 1989.
5. B. Fine, A. M. Gaglione, A. Myasnikov, G. Rosenberger, and D. Spellman, *A classification of fully residually free groups*, J. Algebra (1994) (accepted).
6. R. I. Grigorchuk and P. F. Kurchanov, *Some questions of group theory connected with geometry*, Itogi Nauki i Techniki, Sovremennye problemy matematiki. Fundamental'nye napravlenia (1990); English transl. in Encyclopedia of mathematics.
7. ______, *On quadratic equations in free groups*, Contemp. Math. **131** (1992), no. 1, 159–171.
8. V. Guba, *Equivalence of infinite systems of equations in free groups and semigroups to finite subsystems*, Mat. Zametki **40** (1986), no. 3, 321–324.
9. R. C. Lyndon, *Groups with parametric exponents*, Trans. Amer. Math. Soc. **96** (1960), 518–533.
10. ______, *Equations in groups*, Bol. Soc. Brasil Mat. (N.S.) **11** (1980), 79–102.
11. Roger C. Lyndon and Paul E. Schupp, *Combinatorial group theory*, Springer, 1977.
12. G. S. Makanin, *Equations in a free group*, (Russian) Izv. Akad. Nauk SSSR Ser. Mat. **46** (1982), no. 6, 1199–1273.
13. A. G. Myasnikov and V. N. Remeslennikov, *Exponential groups.* II. *Extensions of centralizers and tensor completion of* CSA-*groups*, Internat. J. Algebra Comput. **6** (1996), no. 6, 687–711.
14. A. Razborov, *On systems of equations in a free group*, Math. USSR, Izvestiya **25** (1985), no. 1, 115–162.
15. ______, *On systems of equations in a free group*, Ph.D. thesis, Steklov Math. Institute, Moscow, 1987.
16. ______, *On systems of equations in free groups*, Combinatorial and Geometric Group Theory (Edinburgh 1993), Cambridge University Press, 1995, pp. 269–283.
17. V. N. Remeslennikov, *∃-free groups and groups with length function*, Second International Conference on Algebra (Barnaul, 1991), 369–376, Contemp. Math. **184**, Amer. Math. Soc., Providence, RI, 1995.
18. ______, *∃-free groups*, Siberian Math. J. **30** (1989), no. 6, 153–157.
19. ______, *∃-free groups as groups with length function*, Ukrainskii Math. J. **44** (1992), no. 3, 813–818.
20. J. R. Stallings, *Fineteness of matrix representation*, Ann. Math. **124** (1986), 337–346.

Department of Mathematics and Statistics, McGill University, 805, Sherbrooke W., Montréal, Québec, H3A 2K6

E-mail address: olga@math.mcgill.ca

Department of Mathematics, City College of City University, Convent Ave & 138th St., New York, N.Y. 10031–9100, USA

E-mail address: alexei@rio.sci.ccny.cuny.edu

Centre de Recherches Mathématiques
CRM Proceedings and Lecture Notes
Volume **17**, 1999

Homological Methods in Group Theory

Yuri Kuz′min

1. Introduction

Some mathematicians who work in combinatorial group theory do not like homological algebra. I think there are two main reasons for this. First of all, to learn homology theory one has to get accustomed to its formalism, rather specific and sometimes tiresome. On the other hand, people are not sure that finally all this machinery would help them to prove theorems in their own field. I hope to convince the reader that homological group theory is not too complicated and is really useful.

These notes are not an introduction to the subject. It's just a guide for the beginners. Some important topics are not even mentioned (for example, Euler characteristic and Poincare duality groups). We concentrate our attention on applications in group theory, giving examples when homological methods work as a tool. The choice of the material was partially determined by the preference of the author. Primarily we consider extensions with abelian kernel, in particular, central extensions. Almost nothing is proved, but we try to explain the reasons and sometimes to indicate the proof.

There are several books where the reader can find a systematic approach or additional information (see for example [**1, 2, 7, 8, 22**]).

2. Definition and Main Properties

Initial data. Let G be an arbitrary group, $\mathbb{Z}G$ its group ring, and A a module over $\mathbb{Z}G$. We are going to define two families of abelian groups

$$H_n(G,A), \quad H^n(G,A) \qquad (n = 0, 1, \dots)$$

called homology and cohomology groups of G with coefficients in A. Both right and left modules will be used in the definition. Fortunately, dealing with group rings, one can make no difference between left and right. Indeed, if A is a right module, then it is also a left module via the action

$$ga = ag^{-1} \quad (a \in A, g \in G).$$

1991 *Mathematics Subject Classification.* Primary: 20J05, 20E22.
The author thanks RFFR (95-01-00733) and INTAS (4373) for financial support.
This is the final form of the paper.

For example, suppose that C and A are two right $\mathbb{Z}G$-modules. Then we can define their tensor product $C \otimes_G A$ factorizing $C \otimes A$, the tensor product over $\mathbb{Z}$, by the relations

$$cg \otimes a = c \otimes ga \quad (a \in A, c \in C, g \in G, ga = ag^{-1}).$$

Exact sequences. Consider a sequence of $\mathbb{Z}G$-modules and homomorphisms

$$\cdots \to A' \xrightarrow{\alpha} A \xrightarrow{\beta} A'' \to \cdots .$$

It is called *exact in* A if $\operatorname{Im} \alpha = \operatorname{Ker} \beta$. For example,

$$0 \to A \xrightarrow{\beta} A''$$

is exact in A if β is a monomorphism, and

$$A' \xrightarrow{\alpha} A \to 0$$

is exact in A if α is an epimorphism. The sequence

$$(1) \qquad 0 \to A' \xrightarrow{\alpha} A \xrightarrow{\beta} A'' \to 0$$

is exact in A', A, and A'' if α is a monomorphism, β is an epimorphism, and $\operatorname{Im} \alpha = \operatorname{Ker} \beta$. In this case (1) is called a short exact sequence. Tensoring by a $\mathbb{Z}G$-module C we get a sequence of abelian groups

$$(2) \qquad 0 \to A' \otimes_G C \to A \otimes_G C \to A'' \otimes_G C \to 0.$$

It can be proved that (2) is exact in $A \otimes_G C$ and $A'' \otimes_G C$. On the other hand, it is not necessarily exact in $A' \otimes_G C$. The map $A' \otimes_G C \to A \otimes_G C$ need not be a monomorphism. If $C = \mathbb{Z}G$ then (1) and (2) are just isomorphic. It can be also verified that if C is free over $\mathbb{Z}G$ then (2) is exact whenever (1) is exact.

Free resolutions. Any $\mathbb{Z}G$-module C is an epimorphic image of a free $\mathbb{Z}G$-module X_0. Equivalently, the sequence

$$X_0 \xrightarrow{d_0} C \to 0$$

is exact. Let C_0 be the kernel of d_0. Then again C_0 is an epimorphic image of a free $\mathbb{Z}G$-module, say X_1. It follows that there is an exact sequence

$$X_1 \xrightarrow{d_1} X_0 \xrightarrow{d_0} C \to 0.$$

Next, one can find a free module X_2 and an epimorphism

$$X_2 \to C_1 = \operatorname{Ker} d_1.$$

Continuing in this way, we see that for any $\mathbb{Z}G$-module C there exists an exact sequence

$$\cdots \xrightarrow{d_{n+1}} X_n \xrightarrow{d_n} \cdots X_1 \xrightarrow{d_1} X_0 \xrightarrow{d_0} C \to 0$$

where all X_n are free $\mathbb{Z}G$-modules. It is called *a free resolution of module* C. The homomorphisms d_n are called *differentials*.

Definition of (co)homology. Let C be an abelian group. Then C can be turned into a $\mathbb{Z}G$-module in a trivial way: $cg = c$ ($c \in C$, $g \in G$). In particular, the additive group of integers $\mathbb{Z}$ can be regarded as a trivial $\mathbb{Z}G$-module. Choose a free resolution of $\mathbb{Z}$

$$\text{(3)} \qquad \cdots \xrightarrow{d_{n+1}} X_n \xrightarrow{d_n} \dots X_1 \xrightarrow{d_1} X_0 \xrightarrow{d_0} \mathbb{Z} \to 0$$

and let A be a $\mathbb{Z}G$-module. To define *homology groups* $H_n(G, A)$, consider the sequence

$$\text{(4)} \qquad \cdots \xrightarrow{\partial_{n+1}} X_n \otimes_G A \xrightarrow{\partial_n} \dots \xrightarrow{\partial_2} X_1 \otimes_G A \xrightarrow{\partial_1} X_0 \otimes_G A \to 0$$

where $\partial_n(x \otimes a) = (d_n x) \otimes a (x \in X_n, a \in A)$. Since $\operatorname{Ker} d_n = \operatorname{Im} d_{n+1}$, it's clear that $\operatorname{Ker} \partial_n \supseteq \operatorname{Im} \partial_{n+1}$. The inclusion can be strict and by definition

$$H_n(G, A) = \operatorname{Ker} \partial_n / \operatorname{Im} \partial_{n+1}.$$

Elements of $\operatorname{Ker} \partial_n$ are called *n-dimensional cycles* and elements of $\operatorname{Im} \partial_{n+1}$ *n-dimensional boundaries.*

To define $H^n(G, A)$, apply $\operatorname{Hom}_G(-, A)$ instead of $\otimes_G A$. Again, the sequence of abelian groups

$$\cdots \xleftarrow{\partial_{n+1}} \operatorname{Hom}_G(X_n, A) \xleftarrow{\partial_n} \dots \xleftarrow{\partial_2} \operatorname{Hom}_G(X_1, A) \xleftarrow{\partial_1} \operatorname{Hom}_G(X_0, A) \leftarrow 0,$$

where $(\partial_n f)(x) = f(d_n x)$ ($x \in X_n$, $f \in \operatorname{Hom}(X_{n-1}, A)$), is not necessarily exact and

$$H^n(G, A) = \operatorname{Ker} \partial_{n+1} / \operatorname{Im} \partial_n.$$

Correctness of the definition. The above definition does not depend on the choice of the free resolution. Let us point out the reason. If X is a free $\mathbb{Z}G$-module with free generators e_i, and a_i are elements of an arbitrary $\mathbb{Z}G$-module X', then the map $e_i \to a_i$ can be extended to a homomorphism $X \to X'$. Similarly, if there is a free resolution $\{X_n\}$ and any exact sequence $\{X'_n\}$ that both start with one and the same module C, then there exists a commutative diagram

$$\begin{array}{ccccccccc}
\cdots \longrightarrow & X_n & \longrightarrow & \cdots \longrightarrow & X_1 & \longrightarrow & X_0 & \longrightarrow & C \\
 & \downarrow & & \downarrow & \downarrow & & \downarrow & & \updownarrow \\
\cdots \longrightarrow & X'_n & \longrightarrow & \cdots \longrightarrow & X'_1 & \longrightarrow & X'_0 & \longrightarrow & C.
\end{array}$$

This fact is known as *comparison theorem.* The proof can be found in any textbook on homological algebra. If also $\{X'_n\}$ is a free resolution, then there exists a similar diagram with vertical arrows in the opposite direction. After tensoring by A over G we still have maps in both directions. They induce isomorphisms of the homology groups.

General properties of the (co)homology groups. Let $f\colon A \to A'$ be a homomorphism of $\mathbb{Z}G$-modules and let (3) be a free resolution of the trivial module $\mathbb{Z}$. Then the maps

$$X_n \otimes_G A \to X_n \otimes_G A'$$

induce homomorphisms of the homology groups

$$f_n\colon H_n(G, A) \to H_n(G, A').$$

It means that $H_n(G, -)$ are functors on the category of $\mathbb{Z}G$-modules with values in the category of abelian groups. Here are the main properties of these functors.

PROPERTY 1. *For any $\mathbb{Z}G$-module A*

$$H_0(G, A) = A/\Delta A \tag{5}$$

where Δ is the fundamental ideal of the group ring $\mathbb{Z}G$ (the ideal generated by elements $g-1$ $(g \in G)$).

To prove this fact, consider the augmentation homomorphism $\epsilon\colon \mathbb{Z}G \to \mathbb{Z}$ ($\epsilon(g) = 1$ if $g \in G$). It's clear that $\operatorname{Ker}\epsilon = \Delta$. Choose an epimorphism $d_1\colon X_1 \to \Delta$, where X_1 is a free $\mathbb{Z}G$-module. Then the sequence

$$X_1 \xrightarrow{d_1} \mathbb{Z}G \xrightarrow{\epsilon} \mathbb{Z} \to 0$$

is the origin of a free resolution for $\mathbb{Z}$. To determine $H_0(G, A)$, we can use the sequence

$$X_1 \otimes_G A \xrightarrow{\partial_1} \mathbb{Z}G \otimes_G A \to 0.$$

The group of 0-dimensional cycles is $\mathbb{Z}G \otimes_G A = A$ and the image of ∂_1 is ΔA. This proves (5).

Note that $A/\Delta A$ is the factor module of A by the relations $ga = a$ ($a \in A$, $g \in G$). In other words, $A/\Delta A$ is the largest quotient of A on which G acts trivially.

PROPERTY 2. *If A is a free $\mathbb{Z}G$-module, then $H_n(G, A) = 0$ for any $n > 0$.*

Indeed, the tensor product of any exact sequence and a free module A is an exact sequence. Thus, if A is free, then in (4) $\operatorname{Ker}\partial_n = \operatorname{Im}\partial_{n+1}$ as was to be proved.

PROPERTY 3 (The long exact sequence).

Surely this is the most important property of homology functors. It concerns the following problem. Let

$$0 \to A' \xrightarrow{\alpha} A \xrightarrow{\beta} A'' \to 0 \tag{6}$$

be a short exact sequence of $\mathbb{Z}G$-modules. Suppose that homology groups $H_*(G, A')$ and $H_*(G, A'')$ are known. Is it possible to get some information about $H_*(G, A)$? The sequence

$$0 \to H_n(G, A') \xrightarrow{\alpha_n} H_n(G, A) \xrightarrow{\beta_n} H_n(G, A'')' \to 0$$

is usually not exact. The kernel of α_n can be nontrivial as well as the cokernel of β_n. It turns out that for $n > 0$ there exist homomorphisms

$$\delta_n\colon H_n(G, A'') \to H_{n-1}(G, A'),$$

called *connected homomorphisms*, such that the sequence

$$0 \to \operatorname{Im}\delta_{n+1} \to H_n(G, A') \xrightarrow{\alpha_n} H_n(G, A) \xrightarrow{\beta_n} H_n(G, A'') \to \operatorname{Im}\delta_n \to 0$$

is exact (for $n = 0$ replace $\operatorname{Im}\delta_n$ by 0). It is possible to put all these maps together and write down the following long exact sequence

$$\cdots \xrightarrow{\beta_{n+1}} H_{n+1}(G, A'') \xrightarrow{\delta_{n+1}} H_n(G, A') \xrightarrow{\alpha_n} H_n(G, A) \xrightarrow{\beta_n} H_n(G, A'') \xrightarrow{\delta_n} \cdots.$$

Certainly, to make use of it, we must define the connecting homomorphisms δ_n. Consider a free resolution (3) of the trivial module $\mathbb{Z}$ and let $c \in X_n \otimes_G A''$ be a cycle that represents an element $h \in H_n(G, A'')$. Choose $\bar{c} \in X_n \otimes_G A$ such that $\bar{c}$ maps to c under the epimorphism $X_n \otimes_G A \to X_n \otimes_G A''$. The image $\partial_n(\bar{c})$ of $\bar{c}$ in $X_{n-1} \otimes_G A$ is not necessarily 0. At the same time $\partial_n(\bar{c})$ is a cycle because

$\operatorname{Im}\partial_n \subseteq \operatorname{Ker}\partial_{n-1}$. Since c is a cycle in $X_n \otimes_G A''$, it's clear that $\partial_n(\bar{c}) = 0$ modulo the image of $X_{n-1} \otimes_G A'$ in $X_{n-1} \otimes_G A$. Actually, we can identify $X_{n-1} \otimes_G A'$ with its image in $X_{n-1} \otimes_G A$ as X_{n-1} is free and therefore the sequence

$$0 \to X_{n-1} \otimes_G A' \to X_{n-1} \otimes_G A X_{n-1} \otimes_G A'' \to 0$$

is exact. Thus $\partial_n(\bar{c})$ is a cycle in $X_{n-1} \otimes_G A'$. We define $\delta_n(h)$ to be the element of $H_{n-1}(G, A')$ that corresponds to $\partial_n(\bar{c})$. Of course, it's necessary to check that δ_n does not depend on the choice of representatives and that the long exact sequence is indeed exact. These facts can be proved rather easily. One should just understand what is given and what is to be proved. This is an amazing feature of homological algebra. When all the objects and arrows are defined in a proper way, everything becomes almost trivial. There is even such a joke.

EXERCISE. Take any book on homological algebra and prove all the theorems not peeping into the proofs given in the book. We leave the details to the reader.

Cohomology functors $H^n(G, A)$ have properties similar, or better to say dual, to Properties 1, 2, 3. The group $H^0(G, A)$ is the largest submodule of A on which G acts trivialy. For $n > 0$ $H^n(G, A) = 0$ if A is a coinduced module, that is, $A = \operatorname{Hom}_{\mathbb{Z}}(\mathbb{Z}G, C)$ where C is an abelian group and

$$(fg)(r) = f(gr) \quad (g \in G, r \in \mathbb{Z}G, f\colon \mathbb{Z}G \to C).$$

Such modules are dual to induced modules $A = C \otimes \mathbb{Z}G$ with the action

$$(c \otimes g)r = c \otimes gr \quad (g \in G, r \in \mathbb{Z}G, c \in C).$$

If C is free abelian, then the induced module A is free and $H_n(G, A) = 0$. Replacing $\otimes$ by Hom we get the notion of coinduced modules which have trivial cohomology.

And the third property. For any short exact sequence (6) there exists the long exact sequence

$$\cdots \xleftarrow{\beta^*_{n+1}} H^{n+1}(G, A'') \xleftarrow{\delta^*_n} H^n(G, A') \xleftarrow{\alpha^*_n} H^n(G, A) \xleftarrow{\beta^*_n} H^n(G, A'') \xleftarrow{\delta^*_{n-1}} \cdots.$$

EXERCISE (not a joke). Define the connecting homomorphisms δ^*_n.

Why is it reasonable? The definition of homology functors is rather complicated. Lecturing on this topic, it is not so easy to convince the audience that we are on the right way. One has quite similar difficulties with determinants in a course for the first-year students. A usual excuse is that determinants have nice applications. Besides, it is the only function $\mathbb{R}^n \to \mathbb{R}$ that is polylinear, skewsymmetric, and equals 1 on the standard basis. We can say just the same about homology. It is good because it has nice applications (we shall see it later). Besides, $H_*(G, -)$ is the only family of functors from the category of $\mathbb{Z}G$-modules to the category of abelian groups that satisfy properties 1, 2, 3 of the previous section. Indeed, $H_0(G, A)$ is determined by the first property. Let $n > 0$. Fix an epimorphism $\sigma\colon X \to A$ where X is free, put $K = \operatorname{Ker}\sigma$ and write down the exact sequence

$$0 \to K \to X \to A \to 0.$$

Since $H_n(G, X) = 0$ for $n > 0$, the corresponding long exact sequence reduces to

$$\cdots 0 \to H_n(G, A) \xrightarrow{\delta_n} H_{n-1}(G, K) \to 0 \to \cdots$$

$$\cdots 0 \to H_1(G, A) \xrightarrow{\delta_1} H_0(G, K) \to H_0(G, X) \to H_0(G, A) \to 0.$$

In other words,

$$H_1(G,A) \cong \operatorname{Ker}\big(H_0(G,K) \to H_0(G,X)\big),$$
$$H_n(G,A) \cong H_{n-1}(G,K) \quad (n > 1).$$

It means that $H_n(G,-)$ can be defined by induction on n.

Topological interpretation. Let $A = \mathbb{Z}$ be the trivial module. Homology groups $H_n(G,\mathbb{Z})$, usually denoted by H_nG, originally appeared in topology. For any reasonable topological space V, for example a CW-complex, one can define homology groups H_nV. Cycles and boundaries in the definition have geometrical nature and H_nV calculates "the number of linearly independent n-dimensional holes". For any group G there exists a topological space V such that the fundamental group $\pi_1(V)$ is isomorphic to G and the higher homotopy groups $\pi_n(V)$ $(n > 1)$ are trivial. Such a space V is called a classifying space of the group G. Up to homotopy equivalence it depends only on G. It can be proved that $H_nG \cong H_nV$. We shall not use this fact. Still, there is no doubt that real understanding of some aspects of homological group theory is impossible without topology. The reader can find many interesting facts on this topic in the book of K. Brown [**2**].

3. Methods of Calculation

Fix a group G. Usually we cannot calculate homology groups $H_n(G,-)$ once and forever because they depend also on the module of coefficients. The best we can do is construct some more or less convenient free resolution for the trivial $\mathbb{Z}G$-module $\mathbb{Z}$. Then it can be used for the calculation of $H_n(G,A)$. Here are some examples.

Free groups. Let G be free on x_i $(i \in I)$. It is easy to verify that the fundamental ideal Δ is then a free $\mathbb{Z}G$-module with free generators $x_i - 1$. The sequence

$$0 \to \Delta \to \mathbb{Z}G \to \mathbb{Z} \to 0$$

is a free resolution of $\mathbb{Z}$ over $\mathbb{Z}G$. In particular, for any $\mathbb{Z}G$-module A $H_n(G,A) = 0$ if $n > 1$ (G has homological dimension 1).

Groups with one defining relation. Suppose that G is generated by elements x_i $(i \in I)$, and let P be a free $\mathbb{Z}G$-module with free generators e_i that are in one-to-one correspondence with the elements x_i. The map $e_i \to x_i - 1$ can be extended to a homomorphism $d_1 \colon P \to \mathbb{Z}G$. It is clear that the fundamental ideal Δ is generated by the elements $x_i - 1$, hence $\operatorname{Im} d_1 = \Delta$. It follows that the initial terms of free resolution can be written as

$$P \xrightarrow{d_1} \mathbb{Z}G \to \mathbb{Z} \to 0.$$

Assume that G is given by one defining relation $r = 1$ on generators x_i. If the word r is not a proper power, then, as Lyndon proved [**20**], $\operatorname{Ker} d_1$ is a free $\mathbb{Z}G$-module of rank 1. Therefore we can write down the following free resolution

$$0 \to \mathbb{Z}G \xrightarrow{d_2} P \xrightarrow{d_1} \mathbb{Z}G \to \mathbb{Z} \to 0.$$

For those who are familiar with Fox derivatives $\partial\ /\partial x_i$, we mention that d_2 maps the free generator to $\sum_i e_i \partial r/\partial x_i$. We also note that under our assumptions for any $\mathbb{Z}G$-module A and $n > 2$ $H_n(G,A) = 0$ (G has homological dimension 2).

Cyclic group of order k. Let $G = \langle x \mid x^k = 1\rangle$. It is easily seen that the fundamental ideal Δ is then a module with one generator $s = x - 1$ and one defining relation $st = 0$ where $t = 1 + x + \cdots + x^{k-1}$. The following sequence is a free resolution of the trivial module $\mathbb{Z}$

$$\cdots \xrightarrow{d_t} \mathbb{Z}G \xrightarrow{d_s} \mathbb{Z}G \xrightarrow{d_t} \mathbb{Z}G \xrightarrow{d_s} \mathbb{Z}G \to \mathbb{Z} \to 0.$$

We write d_r ($r \in \mathbb{Z}G$) for the homomorphism that is multiplication by r. For an arbitrary large n there exist $\mathbb{Z}G$-modules A such that $H_n(G, A) \neq 0$ (G has infinite homological dimension).

EXERCISE. Prove that H_nG is cyclic of order k if n is odd and 0 if n is even ($n > 0$).

Free abelian groups. Let G be a free abelian group with free generators x_1, $\dots$, x_k. Write G in multiplicative form and consider also its isomorphic copy C that is written in additive form. Generators of C that correspond to x_i will be denoted by e_i. Consider the n-th exterior power

$$\Lambda^n = \Lambda^n(C) = C \wedge \cdots \wedge C \quad (n \text{ times}).$$

Recall that to get $\Lambda^n(C)$ from the n-th tensor power $C \otimes \cdots \otimes C$ one must add the relations that make the product $c_1 \otimes \cdots \otimes c_n$ skewsymmetric. Since Λ^n is free abelian on $e_{i_1} \wedge \cdots \wedge e_{i_n}$ ($i_1 < \cdots < i_n$), these elements are free generators of $\mathbb{Z}G$-module $\Lambda^n \otimes \mathbb{Z}G$. The map

$$e_{i_1} \wedge \cdots \wedge e_{i_n} \otimes 1 \to \sum_j (-1)^{n-j} e_{i_1} \wedge \cdots \wedge \hat{e}_{i_j} \wedge \cdots \wedge e_{i_n} \otimes (x_{i_j} - 1),$$

where $\hat{e}_{i_j}$ means the omission of e_{i_j}, defines a homomorphism

$$d_n \colon \Lambda^n \otimes \mathbb{Z}G \to \Lambda^{n-1} \otimes \mathbb{Z}G.$$

It can be proved that the following sequence is a free resolution of $\mathbb{Z}$ over $\mathbb{Z}G$

$$\xrightarrow{d_{n+1}} \Lambda^n \otimes \mathbb{Z}G \xrightarrow{d_n} \dots \Lambda^1 \otimes \mathbb{Z}G \xrightarrow{d_1} \mathbb{Z}G \to \mathbb{Z} \to 0.$$

If k is the number of free generators, then for $n > k$ $\Lambda^n = 0$ and for any $\mathbb{Z}G$-module A $H_n(G, A) = 0$ (G has homological dimension k).

EXERCISE. Show that $H_nG \cong \Lambda^n$.

Group-theoretical constructions. Suppose that G is a free product, free product with amalgamation, or HNN-extension (see [**21**]). Then it is possible to calculate the homology of G under the assumption that we know the homology of the ingredients. For example, let $G = G_1 *_K G_2$ be a free product with amalgamated subgroup K and let A be a $\mathbb{Z}G$-module. Then there is an exact sequence

$$\cdots \xrightarrow{\delta_{n+1}} H_n(K, A) \xrightarrow{\alpha_n} H_n(G_1, A) \oplus H_n(G_2, A) \xrightarrow{\beta_n} H_n(G, A) \xrightarrow{\delta_n} H_{n-1}(K, A) \cdots.$$

Here β_n is the sum of the maps induced by inclusions $G_i \to G$ ($i = 1, 2$) and α_n is the difference of the maps induced by inclusions $K \to G_i$ ($i = 1, 2$). If $K = 1$, then $G = G_1 * G_2$ is a free product. In this case for $n \geq 2$

$$H_n(G, A) \cong H_n(G_1, A) \oplus H_n(G_2, A).$$

EXERCISE. Write down the formula for $H_1(G, A)$.

It is also possible to give a method of calculating of $H_n(G, A)$ where $G = G_1 \times G_2$ is a direct product. For example, if the groups H_pG_1 and H_qG_2 are torsion-free, then

$$H_nG = \sum_{p+q=n} H_pG_1 \otimes H_qG_2.$$

Now let G be an extension of B by N, that is, $N \triangleleft G$ and $B = G/N$. An adequate way to calculate $H_n(G, A)$ is to use a *spectral sequence.* It starts with the groups

$$E^2_{pq} = H_p\big(B, H_q(N, A)\big)$$

where the action of B on $H_q(N, A)$ is induced by conjugation. Then there come successive approximations E^r_{pq} $(r = 2, 3, \dots)$. They are defined in terms of some factors of the groups E^2_{pq}. Spectral sequence is a powerful method but it's not so easy to get the final answer using this method. In general, even for a semidirect product $G = BN$ the situation is very complicated.

Let F be a free group. Denote by $F^{(l)}$ its l-th commutator subgroup and let $G_l = F/F^{(l)}$ be the free solvable group of length l. Then G_l is an extension of G_{l-1} (free abelianized extension of G_{l-1}). The homology groups H_nG_l were studied in [**18**] and [**11**]. It turned out that it is possible to describe H_nG_l using the spectral sequence and induction on l.

A free nilpotent group of class c is the quotient of a free group F by $\gamma_{c+1}(F)$, $(c+1)$-th term of the lower central series. It is a central extension of the free nilpotent group of class $(c-1)$. Very little is known about the spectral sequence of this extension and the homology of free nilpotent groups with integer coefficients. This is one of the actual problems of homological group theory.

Standard resolution. Suppose that G is an arbitrary group. Put $X_0 = \mathbb{Z}G$ and for $n > 0$ let X_n be the free module freely generated by n-tuples $(g_1, \dots, g_n)$ $(g_i \in G,\ g_i \neq 1)$. Consider the sequence

$$\text{(7)} \qquad \cdots \xrightarrow{d_{n+1}} X_n \xrightarrow{d_n} \cdots \xrightarrow{d_2} X_1 \xrightarrow{d_1} X_0 \xrightarrow{d_0} \mathbb{Z} \to 0$$

where $d_0 = \epsilon$ is the augmentation homomorphism, $d_1(g) = g - 1$, and for $n \geq 2$

$$d_n(g_1, \dots, g_n) = (g_1, \dots, g_{n-1})g_n + \sum_{i=n-1}^{1} (-1)^{n-i}(g_1, \dots, g_ig_{i+1}, \dots, g_n) + (-1)^n(g_2, \dots, g_n).$$

For example,

$$d_2(g_1, g_2) = (g_1)g_2 - (g_1g_2) + (g_2)$$
$$d_3(g_1, g_2, g_3) = (g_1, g_2)g_3 - (g_1, g_2g_3) + (g_1g_2, g_3) - (g_2, g_3).$$

It can be easily verified that $\operatorname{Im} d_n \subseteq \operatorname{Ker} d_{n-1}$ (do it for $n = 2,\ 3$). A more tricky argument shows that $\operatorname{Im} d_n = \operatorname{Ker} d_{n-1}$ and hence (7) is a free resolution. It is called the standard resolution of G.

Let $\Delta_n = \Delta \otimes \cdots \otimes \Delta$ be the n-th power of the augmentation ideal and $Y_n = \Delta_n \otimes \mathbb{Z}G$. Then Y_n is free on the tensor products $(g_1 - 1) \otimes \cdots \otimes (g_n - 1)$ $(g_i \in G,\ g_i \neq 1)$. It is possible to use Y_n to define a modification of the standard resolution. Differentials are given by similar formulas.

The standard resolution is not so convenient for calculations, but it is often useful in general arguments.

4. Interpretation of $H^1(G,-)$ and $H^2(G,-)$

Let $\overline{G}$ be an extension of G by an abelian group A. This data can be described as the exact sequence

$$E\colon 1 \to A \to \overline{G} \to G \to 1. \tag{8}$$

The normal subgroup A is called the kernel of the extension E.

Many applications of homological algebra in group theory are, in one way or another, connected with such extensions. For an element $g \in G$ denote by $\bar{g}$ a preimage of g in $\overline{G}$. Since A is abelian, $\bar{g}^{-1}a\bar{g}$ does not depend on the choice of $\bar{g}$ and we can define an action of G on A putting $a^g = \bar{g}^{-1}a\bar{g}$. This action turns A into a $\mathbb{Z}G$-module usually written in additive form. Thus, homology and cohomology groups of G with coefficients in A are defined. We are going to give an interpretation of $H^n(G,A)$ for $n = 1$, 2.

Automorphisms of an extension and $H^1(G,A)$. By definition an automorphism of an extension E is an automorphism $\varphi\colon \overline{G} \to \overline{G}$ that induces the identity map on A and G. Let $\operatorname{Aut} E$ denote the group of all such automorphisms. It is sometimes called the Galois group of the extension E. It's clear that $\operatorname{Aut} E$ contains $\operatorname{Inn}_A E$, the group of inner automorphisms that correspond to elements of A. The quotient $\operatorname{Aut} E/\operatorname{Inn}_A E$ is isomorphic to $H^1(G,A)$. Here is a sketch of the proof. We recall that in the standard resolution

$$d_1(g) = g - 1, \quad d_2(g_1,g_2) = (g_1)g_2 - (g_1g_2) + (g_2).$$

It follows that $H^1(G,A)$ is isomorphic to the group of functions $f\colon G \to A$ that satisfy the condition

$$f(g_1)g_2 - f(g_1g_2) + f(g_2) = 0 \tag{9}$$

modulo the functions α_a such that $\alpha_a(g) = a(g-1)$. Let f be a function that satisfies (9). Then the map $\varphi\colon \overline{G} \to \overline{G}$ that sends any element $\bar{g} \in \overline{G}$ to $\bar{g}f(g)$ is an automorphism (we denoted by g the image of $\bar{g}$ in G). If $f = \alpha_a$, then φ is the inner automorphism and it is easy to show that the correspondence $f \to \varphi$ defines the desired isomorphism.

EXAMPLE 1. Let G_l be free solvable of length l and let $S(G_l)$ be the group of automorphisms of G_l that induce identity map in the factors of its derived series. Shmelkin proved [**25**] that $S(G_l)$ is the group of inner automorphisms defined by elements $a \in G_l^{(l-1)}$, the last nontrivial term of the derived series. Let E be the extension

$$E\colon 1 \to G_l^{(l-1)} \to G_l \to G_{l-1} \to 1.$$

It is proved first that $S(G_l) = \operatorname{Aut} E$. After this the above statement turns into the equality $H^1(G_l, G_l^{(l-1)}) = 0$.

The first cohomology group is often used in the following situation. Suppose that $\overline{G} = GA$ is a semidirect product,in other words, E is a split extension. The complementary subgroup G is usually not unique. If $\varphi \in \operatorname{Aut} E$, then $\varphi(G)$ is also a complement. On the contrary, if $\overline{G} = HA$, $H \cap A = 1$, then $H \cong \overline{G}/A \cong G$ and this isomorphism can be extended to an automorphism of $\overline{G}$ that is the identity map on A. It means that elements of $H^1(G,A)$ are in one-to-one correspondence with the conjugacy classes of the complements of A in G. If $H^1(G,A) = 0$, then all the complements of A in the semidirect product GA are conjugate.

Extentions and $H^2(G, A)$. Consider two extentions

$$E_i\colon 1 \to A \to \overline{G}_i \to G \to 1 \quad (i = 1, 2)$$

with one and the same kernel A and quotient G. By definition, E_1 and E_2 are equivalent if there is an isomorphism $\varphi\colon G_1 \to G_2$ that is identity on A and G. Denote by $E(G, A)$ the set of equivalence classes. We are going to show that there is a bijection

$$E(G, A) \to H^2(G, A),$$

hence $H^2(G, A)$ classifies extensions. Consider extension (8). For any element $g \in G$ choose a preimage $\bar{g} \in G$. If $h \in G$ then $h = \bar{g}a$ ($a \in A$). Since such a presentation is unique, $\overline{G}$ can be identified with the set of pairs (g, a) ($g \in G$, $a \in A$). This suggests that if we want to get all the extensions of G by A, then it is natural to use the set of pairs as building material. Multiplication should be something like

$$(g_1, a_1)(g_2, a_2) = \bigl(g_1g_2, a_1a_2f(a_1, a_2)\bigr)$$

where $f(a_1, a_2) \in A$. The function $f\colon G \times G \to A$ cannot be arbitrary. Associative law gives a restriction on f. A direct calculation shows that $(\bar{g}_1\bar{g}_2)\bar{g}_3 = \bar{g}_1(\bar{g}_2\bar{g}_3)$ if and only if

$$f(g_1, g_2)g_3 - f(g_1, g_2g_3) + f(g_1g_2, g_3) - f(g_2, g_3) = 0.$$

We remind the reader that in the standard resolution

$$d_3(g_1, g_2, g_3) = (g_1, g_2)g_3 - (g_1, g_2g_3) + (g_1g_2, g_3) - (g_2, g_3).$$

It follows that f is a 2-cocycle. Consider another set of representatives $\tilde{g} = \bar{g}\psi(g)$, where $\psi(g) \in A$, and let h be the corresponding 2-cocycle. It is easy to check that f and h define one and the same element of $H^2(G, A)$ (their difference is the (co)boundery $\partial_1(\psi)$ of the function $\psi\colon G \to A$). The map $E \to f$ we have just constructed induces the required bijection. The details are left to the reader.

5. Applications of $H^1(G, -)$ and $H^2(G, -)$

EXAMPLE 2. Let G be a finite group, $N \lhd G$, $B = G/N$. If the orders of the groups N and B are coprime, then the extension

$$1 \to N \to G \to B \to 1$$

splits, and all the complements of N in G are conjugate.

Several words about the proof of this classical result. Suppose first that $N = A$ is an abelian normal subgroup. The exponents of the cohomology groups $H^n(B, A)$ divide the exponent of A. Indeed, $mA = 0$ implies $mf = 0$ for any homomorphism with values in A. A beautiful calculation shows that for $n > 0$ the exponents of $H^n(B, A)$ also divide the order of B [**22**, Ch. 4, Sec. 5]. If the orders of A and B are coprime, then it follows that for $n > 0$ $H^n(B, A) = 0$. In particular, $H^1(B, A) = H^2(B, A) = 0$. This exactly means that the extension splits and all the complements are conjugate. To prove the first part of the theorem in the general case, one should use induction on the order, Sylow subgroups and their normalizers [**22**, Ch. 4, Sec. 10].

The first part of the theorem was proved by Schur. For a solvable normal subgroup N the second part was proved by Zassenhaus [**30**] (it can be easily reduced to the abelian case by induction on the length of solvability). Chunihin [**4**] proved the second part for the case when B is solvable. But by Thompson-Feit theorem any group of odd order is solvable, so either N or B is solvable.

EXAMPLE 3. If in the extension

$$1 \to A \to G \to B \to 1$$

the kernel A is a vector space over the field of rational numbers $\mathbb{Q}$ and B is finite, then such extension splits and all the complements are conjugate.

Indeed, as we mentioned before, $H^n(B,-)$ $(n > 0)$ has finite exponent that divides the order of B. On the other hand, $H^n(B, A)$ is a vector space over $\mathbb{Q}$. The only possibility is that $H^n(B, A) = 0$. In particular, $H^2(B, A) = H^1(B, A) = 0$, which proves the statement.

Proceeding by induction on the length of solvability it is easy to prove that the same is true if A has a series of subgroups A_i normal in G

$$A = A_1 \supseteq A_2 \supseteq \cdots \supseteq A_l = 0$$

such that A_i/A_{i-1} are vector spaces over $\mathbb{Q}$.

EXAMPLE 4. A group G is called a $\mathbb{Q}$-group if for any element $g \in G$ and integer n there is a unique element $h \in G$ such that $g = h^n$. If $\lambda = m/n \in \mathbb{Q}$ then it is natural to put $g^\lambda = h^m$ (this explains the term $\mathbb{Q}$-group). An abelian $\mathbb{Q}$-group written in additive form is evidently a vector space over $\mathbb{Q}$. If G is nilpotent and torsion-free, then G can be embedded in a unique $\mathbb{Q}$-group G^* such that for any element $g \in G^*$some power of g belongs to G. This $\mathbb{Q}$-group G^* is called the Maltcev completion of G. It is nilpotent of the same class as G.

Let G be a free nilpotent group of class c with free generators $x_1, \ldots, x_n$. Then G^* is a free nilpotent $\mathbb{Q}$-group with free generators $x_1, \ldots, x_n$. For any elements $y_1, \ldots, y_n \in G^*$ the map $x_i \to y_i$ can be extended to an endomorphism $\varphi\colon G^* \to G^*$. It is an automorphism if and only if φ induces an automorphism $\varphi_{ab}\colon G^*_{ab} \to G^*_{ab}$ of the n-dimensional vector space G^*_{ab}. It follows that any single automorphism of G^*_{ab} can be lifted to an automorphism of G^*. The map $\varphi \to \varphi_{ab}$ is an epimorphism

$$\Phi\colon \operatorname{Aut} G^* \to \mathrm{GL}_n(\mathbb{Q}).$$

Let B be a finite subgroup of $\mathrm{GL}_n(\mathbb{Q})$. Can it be isomorphically lifted to a finite subgroup $\overline{B}$ of $\operatorname{Aut} G^*$? The answer is positive; moreover, if $\overline{B}_i \subseteq \operatorname{Aut} G^*$ $(i = 1, 2)$ are two subgroups such that Φ induces isomorphisms $\overline{B}_i \to B$ $(i = 1, 2)$, then $\overline{B}_1$ and $\overline{B}_2$ are conjugate in $\operatorname{Aut} G^*$ [**13**]. We just outline the proof. For any automorphism $\varphi \in \operatorname{Aut} G^*$ denote by φ_k the automorphism that φ induces in $G^*/\gamma_{k+1}(G^*)$, the factor group by the $(k+1)$-th term of the lower central series. Let A_k be the subgroup of $\operatorname{Aut} G^*$ that consists of automorphisms φ such that $\varphi_k = id$. It can be easily verified that the factors of the series

$$\operatorname{Ker} \Phi = A_1 \supseteq A_2 \supseteq \cdots \supseteq A_c = 1$$

are abelian $\mathbb{Q}$-groups, hence it is possible to use the previous example.

EXAMPLE 5. If any extension of G splits, then G is a free group.

This theorem was proved by Stallings for finitely generated groups [**27**] and then by Swan in the general case [**29**]. It is sufficient to assume that only extensions with abelian kernel split. In other words , it is sufficient to prove that if for any $\mathbb{Z}G$-module A $H^2(G, A) = 0$, then G is free. Since extensions of free groups split, this theorem gives a characterization of groups of cohomological dimension 1. Probably it's time to give a formal definition. The cohomological dimension of G is the least number n such that for any $\mathbb{Z}G$-module A $H^{n+1}(G, A) = 0$.

EXAMPLE 6. If a torsion-free group G contains a free subgroup F of a finite index, then G itself is free.

Note that in the statement of this result homology and extensions are not even mentioned. Serre proved that under assumptions of the theorem $H^2(G, A) = 0$ for any $\mathbb{Z}G$-module A. Then by the theorem of Stallings and Swan G is free.

EXAMPLE 7. It can be easily proved that the cohomological dimension of a subgroup is less than or equal to the cohomological dimension of the whole group. In particular, if G is a subgroup of a free group, then G has cohomological dimension 1 and therefore is free by the above theorem. Thus, there is a homological method to prove that a subgroup of a free group is free. Of course, it is easier to give a combinatorial proof, but there are categories in which this trick invented by Tate is natural. We recall that a pro-p-group is an inverse limit of finite p-groups. One can develop cohomology theory for pro-p-groups and prove that a pro-p-group of cohomological dimension 1 are free in the category of pro-p-groups. The proof is much easier than for groups [**24**]. It follows that a closed subgroup of a free pro-p-group is also a free pro-p-group. A combinatorial argument is scarcely possible.

A similar theorem can be proved "in characteristic 0". To be more exact, we say that a $\mathbb{Q}$-group has finite dimension if it has a normal series of finite length such that its factors are isomorphic to the additive group of rationals. It can be proved that $\mathbb{Q}$-groups of finite dimension are nipotent [**3**] (just as finite p-groups). By definition, a pro-$\mathbb{Q}$-group is an inverse limit of $\mathbb{Q}$-groups of finite dimension. There exists appropriate cohomology theory for pro-$\mathbb{Q}$-groups; pro-$\mathbb{Q}$-groups of cohomological dimension 1 are free in the category of pro-$\mathbb{Q}$-groups, and a closed subgroup of a free pro-$\mathbb{Q}$-group is also free [**17**].

EXAMPLE 8. Let $\overline{G} = GA$ be a semidirect product (A is an abelian normal subgroup). If $\varphi \in \operatorname{End}_{\mathbb{Z}G} A$, the ring of endomorphisms of the $\mathbb{Z}G$-module A, then φ can be extended to an endomorphism $\bar{\varphi}$ of the group $\overline{G}$. One can just put $\bar{\varphi}(ga) = g\varphi(a)$ ($g \in G$, $a \in A$). Suppose now that E is an arbitrary extension with abelian kernel A and again $\varphi \in \operatorname{End}_{\mathbb{Z}G} A$. When is it possible to extend φ to an endomorphism $\bar{\varphi}: \overline{G} \to \overline{G}$ that is identity $\mod A$? The answer can be given in homological terms. Let e be an element of $H^2(G, A)$ that corresponds to E and let φ^* be the induced homomorphism $\varphi^*: H^2(G, A) \to H^2(G, A)$. It turns out that $\bar{\varphi}$ exists if and only if $\varphi^*(e) = e$. It is easy to see that if both $\bar{\varphi}_1$ and $\bar{\varphi}_2$ extend φ, then $\bar{\varphi}_2 = \bar{\varphi}_1\alpha$ where α is identity on A and $\mod A$, that is $\alpha \in \operatorname{Aut} E$. This information can be summarized in the following exact sequence

$$1 \to \operatorname{Aut} E \to \operatorname{End}_G \overline{G} \to \operatorname{St}(e) \to 1,$$

where $\operatorname{End}_G \overline{G}$ is the semigroup of endomorphisms of $\bar{G}$ that induce the identity map in G, and $\operatorname{St}(e)$ is the stabilizer of e in $\operatorname{End}_{\mathbb{Z}G} A$.

Consider the case when G is also abelian. Then for any $r \in \mathbb{Z}G$ multiplication by r defines an endomorphism φ_r of $\mathbb{Z}G$-module A. For $r = g \in G$ φ_r is the inner automorphism defined by g. It is known that inner automorphisms induce the identity map on all the groups $H^n(G, A)$, therefore $\varphi_r^*(e) = e^{\epsilon(r)}$ where ϵ is the augmentation map $\epsilon\colon \mathbb{Z}G \to \mathbb{Z}$. It follows that for any $r \in \mathbb{Z}G$ such that $\epsilon(r) = 1$ φ_r can be extended from A to $\overline{G}$.

EXERCISE. Prove that for $r = 1 + \sum n_i(g_i - 1)$ ($g_i \in G$, $n_i \in \mathbb{Z}$) the map $\bar{\varphi}_r$ can be defined by the formula

$$\bar{\varphi}_r(h) = h \prod [h, \bar{g}_i]^{n_i}$$

where $h \in \overline{G}$, $\bar{g}_i$ are preimages of g_i in $\bar{G}$. The details can be found in [**14**].

6. Interpretation of H_1G and H_2G

We described some applications of the cohomology groups $H^1(G, A)$, $H^2(G, A)$. The second homology group with integer coefficients H_2G is also important for group theory. It is connected with defining relations and central extensions. To calculate the groups H_nG, one can use the resolution (Y_n, d_n), which is a modification of the standard resolution. We remind the reader that $Y_n = \Delta_n \otimes \mathbb{Z}G$, where $\Delta_n = \Delta \otimes \cdots \otimes \Delta$ is the n-th tensor power of the augmentation ideal. After tensoring by $\mathbb{Z}$ over $\mathbb{Z}G$ we get the complex (Δ_n, ∂_n) with differentials given by the formula

$$\partial_n(s_1 \otimes \cdots \otimes s_n) = \sum_{i=n}^{1} (-1)^{n-i} s_1 \otimes \cdots \otimes s_{i-1}s_i \otimes \cdots \otimes s_n \quad (s_i \in \Delta).$$

For example,

$$\partial_1(s_1) = 0, \quad \partial_2(s_1 \otimes_2) = s_1 s_2,$$
$$\partial_3(s_1 \otimes s_2 \otimes s_3) = s_1 \otimes s_2 s_3 - s_1 s_2 \otimes s_3.$$

Since $H_nG = \operatorname{Ker} \partial_n / \operatorname{Im} \partial_{n+1}$, we have

$$H_1G = \Delta / \langle s_1 s_2 \rangle,$$
$$H_2G = \operatorname{Ker}(\Delta \otimes \Delta \to \Delta) / \langle s_1 \otimes s_2 s_3 - s_1 s_2 \otimes s_3 \rangle$$

or, equivalently,

$$H_1G = \Delta / \Delta^2, \quad H_2G = \operatorname{Ker}(\Delta \otimes_{ZG} \Delta \to \Delta).$$

It is an easy exercise to prove that the map $gG' \to (g - 1) + \Delta^2$ defines an isomorphism $G_{ab} \cong \Delta/\Delta^2$, therefore $H_1G \cong G_{ab}$. A more complicated argument shows that if G is given as a factor group of a free group $G = F/N$, then the kernel of the map $\Delta \otimes_{ZG} \Delta \to \Delta$ is isomorphic to $N \cap F'/[N, F]$ and hence

$$H_2G \cong N \cap F'/[N, F]. \tag{10}$$

This formula is due to Hopf [**9**]. It was one of the first theorems in homological group theory. If $G = F/N$, then elements of the normal subgroup N are relations of G. By (10) any relation $v \in N \cap F'$ defines an element $h(v) \in \operatorname{Ker}(\Delta \otimes_{ZG} \Delta \to \Delta)$. It is possible to give an explicit expression for $h(v)$. If x_i are free generators of F, then

$$h(v) = \sum_i (x_i - 1) \otimes (\partial v / \partial x_i)$$

where $\partial v/\partial x_i$ are the images of Fox derivatives in Δ. Note that

$$\partial_2(h(v)) = \sum_i (x_i - 1)(\partial v / \partial x_i).$$

By the Fox identity, the sum on the right is $v - 1$. This is zero in Δ since v is a relation, thus $h(v)$ is indeed a 2-cycle. It can be proved that if $v \in [N, F]$, then $h(v)$ is a boundary. The isomorphism (10) is induced by the map $v \to h(v)$.

Here is a simple example of 2-dimensional cycles. Suppose that $g_1 g_2 = g_2 g_1$ (g_1, $g_2 \in G$). It is obvious that the element $(g_1, g_2) - (g_2, g_1)$ of the standard resolution is a 2-cycle. It is also clear that $(g_1 - 1) \otimes (g_2 - 1) \in \operatorname{Ker}(\Delta \otimes_{ZG} \Delta \to \Delta)$. If $g = F/N$ and $\bar{g}_1$, $\bar{g}_2$ are preimages of g_1, g_2 in $F/[N, F]$, then under the isomorphism (10) this 2-cycle corresponds to the commutator $[\bar{g}_1, \bar{g}_2]$. The corresponding elements of $H_2 G$ will be denoted by $g_1 \wedge g_2$. In K-theory 2-cycles of this type are called symbols. They are bilinear and skewsymmetric.

7. H_2G and Relations

The following fact gives a necessary condition for a group to be finitely presented. If H_2G is not finitely generated, then G is not finitely presented. Indeed, otherwise $G = F/N$, where F is a finitely generated free group and N is finitely generated as a normal subgroup. It follows that $N/[N, F]$ is a finitely generated abelian group and hence so is its subgroup $N \cap F'/[N, F] \cong H_2G$.

Of course the second homology group does not contain complete information on relations. There are a lot of groups with trivial H_2G which have nontrivial relations. Nevertheless, the equality $H_2G = 0$ can be sometimes useful by itself.

EXAMPLE 9. We recall that an n-knot is a smooth embedding $f\colon S^n \to \mathbb{R}^{n+2}$ of the n-dimensional sphere S^n into the $(n+2)$-dimensional space $\mathbb{R}^{n+2}$.The group G of the knot f is the fundamental group of the complement $\mathbb{R}^{n+2} \backslash f(S^n)$. Let $n \geq 3$. As Kervaire proved [**10**], *a group G is isomorphic to the group of an n-knot if and only if*

1. *G is a normal closure of a single element*;
2. *G_{ab} is infinite cyclic*;
3. $H_2G = 0$.

EXAMPLE 10. We say that G is a C-group if G can be given by a presentation $G = \langle X \mid R \rangle$ in which all the relations are conjugations

$$x_j^{-1} x_i x_j = x_k \quad (x_* \in X).$$

By definition, a C-group is irreducible if G_{ab} is cyclic or, equivalently, the images of the generators are conjugate in G.

A knotted surface is a smooth embedding $f\colon S \to \mathbb{R}^4$ of a compact orientable connected surface (a sphere with handles) into the four-dimensional space. The group of a knotted surface is the fundamental group of the complement $\mathbb{R}^4 \backslash f(S)$. It was proved in [**12**] and [**5**] that G is isomorphic to the group of a knotted surface if and only if G is a finitely presented irreducible C-group. It turns out that such groups can be characterized in terms of the second homology group. It was proved in [**19**] that *a group G is an irreducible C-group if and only if there exists an element $g \in G$ such that*

1. *G is the normal closure of g*;
2. *G_{ab} is infinite cyclic*;
3. *Any element of H_2G is equal to $g \wedge h$ for some $h \in C(g)$ (the centralizer of g)*.

EXAMPLE 11. Let G be a finite p-group. Denote by n the minimal number of generators of G, by r the minimal number of relations between these generators, and by d the dimension of the second homology group $H_2(G, \mathbb{Z}_p)$ with coefficients in the prime field $\mathbb{Z}_p$. It is easy to see that $r \geq d$. As Golod and Shafarevich proved [**6**], $d \geq (1/4)(n-1)^2$. It follows that when the number of generators n tends to infinity the deficiency $r - n$ also tends to infinity. This deep result has consequences in algebraic number theory.

8. H_2G and Central Extensions

An extension

$$E\colon 1 \to A \to \overline{G} \to G \to 1 \tag{11}$$

is called central if A is contained in the center of $\overline{G}$. It means that A is a trivial $\mathbb{Z}G$-module. Conversely, fix a trivial $\mathbb{Z}G$-module A. Then all the elements of the cohomology group $H^2(G, A)$ define central extensions.

Our first remark on central extensions is, that if in (11) G_{ab} and $\overline{G}_{ab}$ are free abelian, then $\overline{G} = C \times \widetilde{G}$ for some subgroups $C \subseteq A, \widetilde{G} \subseteq \overline{G}$, where $\widetilde{G}_{ab} \cong \overline{G}_{ab}$ under the isomorphism induced by the projection $\overline{G} \to \widetilde{G}$. Indeed, the image of the natural homomorphism $\varphi\colon A \to \overline{G}_{ab}$ is a subgroup of the free abelian group $\overline{G}_{ab}$ and therefore is itself free abelian. This implies that the sequence

$$1 \to \operatorname{Ker}\varphi \to A \to \operatorname{Im}\varphi \to 1$$

splits. Let C be a complement of $\operatorname{Ker}\varphi$ in A. By our hypothesis the quotient group $\overline{G}_{ab}/\operatorname{Im}\varphi \cong G_{ab}$ is also free abelian, hence $\overline{G}_{ab} = \operatorname{Im}\varphi \times D$ for some subgroup $D \subseteq \overline{G}_{ab}$. Consider the composition of the natural map $\overline{G} \to \overline{G}_{ab}$, the projection $\overline{G}_{ab} \to \operatorname{Im}\varphi$ and the splitting isomorphism $\operatorname{Im}\varphi \to C$. It follows from the definitions that f is a retraction of $\overline{G}$ on C. If we put $\widetilde{G} = \operatorname{Ker} f$, then clearly $\overline{G} = C \times \widetilde{G}$ and $\widetilde{G}_{ab} \cong D \cong G_{ab}$. Note that the embedding $\widetilde{G} \to \overline{G}$ defines the following commutative diagram

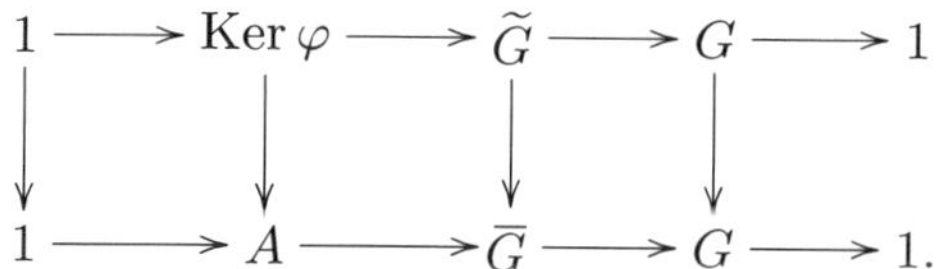

The subgroup $\widetilde{G}$ depends on the choice of C and D, but for any two such subgroups $\widetilde{G}_1$ and $\widetilde{G}_2$ there is an isomorphism $\widetilde{G}_1 \to \widetilde{G}_2$ that is identity on $\operatorname{Ker}\varphi$ and the factor group G. It means that $\widetilde{G}_1$ and $\widetilde{G}_2$ define equivalent central extensions.

Now suppose that G is given as a quotient $G = F/N$ with F a free group and N a normal subgroup of F. The group $\hat{G} = F/[F, N]$ is called the free central extension of G. By the Hopf formula, the kernel of the natural homomorphism $\varphi\colon N/[N, F] \to \hat{G}_{ab} = F_{ab}$ is isomorphic to H_2G. If G_{ab} is a free abelian group, then we can apply the above construction. It gives a central extension

$$1 \to H_2G \to \widetilde{G} \to G \to 1, \tag{12}$$

which is called the universal central extension of G. Note that if $N \subset F'$ then $\widetilde{G} = \hat{G}$. The free central extension $\hat{G} = F/[F, N]$ depends in general on the choice of a presentation $G = F/N$. On the other hand, the corresponding universal extension $\widetilde{G}$ is unique up to an isomorphism that is identity on H_2G and G. It can be proved that for an arbitrary central extension (11) there is a homomorphism $f\colon \widetilde{G} \to \overline{G}$ that is identity on G. If f_1 and f_2 are two such homomorphisms, then $f_2 = \alpha f_1$, where $\alpha\colon \overline{G} \to \overline{G}$ is an automorphism of E. This explains the term universal central extension.

EXERCISE. Prove that if $G_{ab} = 1$ then f is unique.

EXAMPLE 12. Let Λ be an associative ring with 1. For any $\lambda \in \Lambda$ and $i, j \leq n$ ($i \neq j$) denote by $e_{ij}^{\lambda} \in \mathrm{GL}(n, \Lambda)$ the elementary matrix with λ in position (i, j) (in the other positions e_{ij}^{λ} is equal to the identity matrix). It is easily verified that

1. $e_{ij}^{\lambda} e_{ij}^{\mu} = e_{ij}^{\lambda+\mu}$;
2. $[e_{ij}^{\lambda}, e_{jl}^{\mu}] = e_{il}^{\lambda\mu}$ if $i \neq l$;
3. $[e_{ij}^{\lambda}, e_{kl}^{\mu}] = 1$ if $i \neq l$, $j \neq k$.

Let $E(n, \Lambda)$ be the subgroup of $\mathrm{GL}(n, \Lambda)$ generated by all elementary $n \times n$ matrices, and let $E(\Lambda) = \cup_n E(n, \Lambda)$. It is known that $E(\Lambda)_{ab} = 1$. By definition, the Steinberg group $\mathrm{St}(\Lambda)$ is the group given by generators x_{ij}^{λ} ($\lambda \in \Lambda$, $i, j \geq 1$, $i \neq j$) and defining relations 1, 2, 3 in which the elements e_{ij}^{λ} are replaced by x_{ij}^{λ}. It's clear that the map $x_{ij}^{\lambda} \to e_{ij}^{\lambda}$ can be extended to an epimorphism $\mathrm{St}(\Lambda) \to E(\Lambda)$. By definition, the kernel of this map is $K_2\Lambda$. It can be proved that $K_2\Lambda$ is contained in the center of $\mathrm{St}(\Lambda)$. Moreover, $\mathrm{St}(\Lambda)$ is the universal central extension of $E(\Lambda)$, so $K_2\Lambda \cong H_2\big(E(\Lambda)\big)$ [**23**]. The group $K_2\Lambda$ was determined for many rings. For example, $K_2\mathbb{Z}$ is cyclic of order 2.

EXAMPLE 13. Suppose again that $G = F/N$ where F is a free group. As we saw above, the kernel of the free central extension $F/[N, F]$ contains H_2G. It is known that for any finitely generated group A there exists a finitely presented group G such that $H_2G \cong A$. In particular, $F/[N, F]$ can have elements of arbitrary finite order.

The group F/N', where $N' = [N, N]$ is the commutator subgroup of N, is called the free abelianized extension of G. By definition, there is an exact sequence

$$1 \to N_{ab} \to F/N' \to G \to 1.$$

In contrast with the free central extension, F/N' is torsion-free for any group G. Indeed, let $g \in F$. Denote by Φ the subgroup generated by g and N. We want to prove that Φ/N' is torsion free. This group is an extension of Φ/Φ' by Φ'/N'. The first quotient is a free abelian group because Φ is free. The second is also free abelian because it is a subgroup of the free abelian group N/N'.

We are going to consider a combination of these two constructions, the group $F/[N', F]$, which is a free central extension of free abelianized extension F/N'. Since F/N' is torsion-free, elements of finite order belong to the central subgroup $N'/[N', F]$. Which orders are possible? The answer is rather unexpected [**15**]. The only possible finite orders are 2 and 4.

By the Hopf formula, $N'/[N', F] \cong H_2(F/N')$. This group was described [**18**] in terms of $H_3(G, \mathbb{Z}_2), H_4(G, \mathbb{Z}_2)$, and centralizeres of involutions of the group $G = F/N$. For example, if G has no involutions, then [**16, 18, 28**] the torsion subgroup of $H_2(F/N')$ is isomorphic to $H_4(G, \mathbb{Z}_2)$ and only elements of order 2 are possible. This example shows that $H_4(-)$ and $H_3(-)$ also appear in applications.

EXERCISE. What is the torsion subgroup of the free center-by-metabelian group $F/[F'', F]$?

Suppose again that G_{ab} is free abelian and let E be a central extension of G given by (11). If (12) is the universal central extension, then there is a homomorphism $f\colon \widetilde{G} \to \overline{G}$. The restriction of f to H_2G defines an element $e \in \operatorname{Hom}(H_2G, A)$. It turns out that the map $E \to e$ defines an isomorphism

$$H^2(G, A) \cong \operatorname{Hom}(H_2G, A).$$

In this sense H_2G classifies central extensions. In particular, if $H_2G = 0$, then any central extension of G splits. The situation is not much more complicated if G_{ab} is not free abelian. In this case there is a homomorphism

$$H^2(G, A) \to \operatorname{Hom}(H_2G, A).$$

The kernel is isomorphic to the group $\operatorname{Ext}(G_{ab}, A)$ that can be described as the subgroup of $H^2(G_{ab}, A)$ that consists of extensions

$$1 \to A \to B \to G_{ab} \to 1$$

with abelian group B.

In conclusion we consider an example in which defining relations and central extensions are not mentioned.

EXAMPLE 14. Let G be an arbitrary group, and let $\gamma_n(G)$ denote the n-th term of the lower central series of G. Put $L_n = \gamma_n(G)/\gamma_{n+1}(G)$ and consider the direct sum $L(G) = \oplus_n L_n$. If $a \in L_n$, $b \in L_m$ and $\bar{a}$, $\bar{b}$ are their preimages in G, then the equality

$$(a, b) = [\bar{a}, \bar{b}]\gamma_{n+m+1}(G)$$

defines a multiplication $L_n \times L_m \to L_{n+m}$. By linearity it can be extended to $L(G)$. It is well known that under these operations $L(G)$ becomes a Lie ring. A group homomorphism $\varphi\colon G_1 \to G_2$ induces a homomorphism of Lie rings $L(\varphi)\colon L(G_1) \to L(G_2)$. Consider the induced homomorphisms of the homology groups $\varphi_n\colon H_nG_1 \to H_nG_2$. Stallings proved [**27**] that if φ_1 is an isomorphism and φ_2 is an epimorphism, then $L(\varphi)$ is an isomorphism.

References

1. A. Babakhanian, *Cohomology methods in group theory*, Marcel Dekker Inc., 1972.
2. K. S. Brown, *Cohomology of groups*, Grad. Texts in Math., vol. 87, Springer-Verlag, New York-Berlin, 1982.
3. V. Charin, *On complete groups with root series of finite length*, Dokl. Akad. Nauk SSSR **66** (1949), 809–811 (Russian).
4. S. Chunihin, *Subgroups of finite groups*, Minsk, 1964 (Russian).
5. N. Gilbert and J. Howie, LOG *groups, higher knot groups and Schur multipliers*, Bull. London Math. Soc.
6. E. Golod and I. Shafarevich, *On the class field tower*, Izv. Akad. Nauk SSSR Ser. Mat. **28** (1964), 261–272 (Russian).
7. K. W. Gruenberg, *Cohomological topics in group theory*, Lecture Notes in Math., vol. 143, Springer-Verlag, Berlin-New York, 1970.
8. P. Hilton and U. Stammbach, *A course in homological algebra*, Grad. texts in Math., vol. 4, Springer-Verlag, New York-Berlin, 1971.
9. H. Hopf, *Fundamentalgruppe und zweite Bettische Gruppe*, Comment. Math. Helv. **14** (1942), 257–309.
10. M. A. Kervaire, *On higher dimensional knots*, Differential and Combinatorial Topology, Princeton Univ. Press, Princeton, 1965, pp. 105–119.
11. L. G. Kovach, Yu. V. Kuz′min, and R. Shter, *Homology of free Abelian extensions of groups*, Mat. Sb. **182** (1991), no. 4, 526–542; English transl., Math. USSR-Sb. **72** (1992), no. 2, 503–518.

12. V. Kulikov, *Geometric realization of C-groups*, Izv. Ross. Akad. Nauk, Ser. Mat. **58** (1994), no. 4, 194–203; English transl., Russian Acad. Sci. Izv. Math. **45** (1995), no. 1, 197–206.
13. Yu. Kuz′min, *Representations of finite groups by automorphisms of nilpotent groups*, Sibirsk. Mat. Zh. **13** (1972), 107–117 (in Russian).
14. ______, *Inner endomorphisms of metabelian groups*, Sibirsk. Mat. Zh. **16** (1986), 736-744 (Russian).
15. ______, *Finite-order elements in free groups of certain manifolds*, Mat. Sb. (N.S.) **119(161)** (1982), no. 1, 119–131 (Russian); English transl., Math. USSR-Sb. **47** (1984), no. 1, 115–126.
16. ______, *Certain varieties of free groups*, Mat. Sb. (N.S.) **125(167)** (1984), no. 1(9), 128–142 (Russian); English transl., Math. USSR-Sb. **53** (1986), 131–145.
17. ______,*On theorems of the Nielsen-Schreier theorem type in an algebra of series in non-commuting variables*, Sibirsk. Mat. Zh. **27** (1986), no. 2, 91–103 (Russian); English transl., Siberian Math. J. **27** (1986), no. 2, 217–226.
18. ______, *Homology theory of free Abelianized extensions*, Comm. Algebra, **16** (1988), no. 12, 2447–2533.
19. ______, *The groups of knotted compact surfaces, and central extensions*, Mat. Sb. **187** (1996), no. 2, 81–102; English transl., Russian Acad. Sci. Sb. Math. **187** (1996), 237–257.
20. R. C. Lyndon, *Cohomology theory of groups with a single defining relation*, Ann. of Math. (2) **52** (1950), 650–665.
21. R. C. Lyndon and P. E. Schupp, *Combinatorial group theory*, Ergeb. Math. Grenzgeb., vol. 89, Springer-Verlag, Berlin-New York, 1977.
22. S. MacLane, *Homology*, Grundlehren Math. Wiss., vol. 114, Academic Press, New York; Springer-Verlag, Berlin-Göttingen-Heidelberg, 1963.
23. J. Milnor, *Introduction to algebraic K-theory*, Ann. of Math. Stud., vol. 72, Princeton Univ. Press, Princeton; Univ. of Tokyo Press, Tokyo, 1971.
24. J.-P. Serre,*Cohomologie galoisienne*, Lecture Notes in Math., vol. 5, Springer-Verlag, Berlin-New York, 1965.
25. A. Shmelkin, *Two remarks on free solvable groups*, Algebra i Logika, **6** (1967), 95–109 (Russian).
26. J. R. Stallings, *Homology and central series of groups*, J. Algebra **2** (1965), 170–181.
27. ______, *On torsion-free groups with infinitely many ends*, Ann. of Math. (2) **88** (1968), 312–334.
28. R. Stöhr, *On torsion in free central extensions of some torsion-free groups*, J. Pure Appl. Algebra **46** (1987), no. 2-3, 249–289.
29. R. G. Swan, *Groups of cohomological dimension one*, J. Algebra, **12** (1969), 585–610.
30. H. J. Zassenhaus, *The theory of groups*, Chelsea, New York, 1958.

Department of Mathematics, Moscow State University of Railways, Obraztsov St. 15, 101475 Moscow, Russia

E-mail address: m10107@sucemi.bitnet

Centre de Recherches Mathématiques
CRM Proceedings and Lecture Notes
Volume 17, 1999

Pro-p Groups

Marcus du Sautoy

The theory of pro-p groups has undergone something of a renaissance in the last ten years. Part of the reason for this has been the remarkable array of results about abstract groups that have been proved by using pro-p groups. But the world of pro-p groups has become a rich land in its own right for exploration recently. The successful acceptance of p-adic analytic groups into the group theorist's arsenal has emboldened us to explore beyond these realms. There is still much uncharted water but some sort of map is beginning to emerge of this world.

This paper is based on a series of lectures given to the Banff Summer School in Group Theory. The audience at the school ended up being rather mixed—a combination of graduate students and none experienced mathematicians. So this paper will reflect that mix. It is meant as some sort of introduction to the theory of pro-p groups which is on firm ground and well documented together with some of the more uncharted territory.

The paper will concentrate on pro-p groups for their own sake rather than their applications to abstract group theory. But I hope that this will not put off those who may not be directly interested in pro-p groups. It is always worth checking that the group you are exploring may have a pro-p completion or a pro-p analogue which fits nicely into the world we understand and that there is a wealth of results that are there for you to plunder. Alternatively, you might find that this pro-p group that you have constructed is a new island which you can claim as your own.

1. Definitions

Pro-p groups are to profinite groups what finite p-group are to finite groups. So we begin with the definition of a profinite group. There are two equivalent formulations: one as a special sort of topological group; the other a constructive definition which strings finite groups together into a big infinite group. We start with the first:

DEFINITION 1. A profinite group is a compact, Hausdorff topological group whose open subgroups form a base of neighbourhoods of the identity.

Recall that $\{X_i\}_{i\in I}$ is called a base of neighbourhoods of the identity for a group G if whenever U is an open subset of G containing the identity, U must contain one of the subsets X_i for some $i \in I$.

1991 *Mathematics Subject Classification.* Primary: 20E18; Secondary: 22E20.
This is the final form of the paper.

The following proposition records some of the properties of such a profinite topology on a group G:

PROPOSITION 1. (1) *If H is an open subgroup then H has finite index in G.*
(2) *U is an open subset in G if and only if U is a union of cosets of open normal subgroups.*
(3) *The intersection of all open subgroups is just the identity element:*

$$\bigcap_{H \text{ open in } G} H = 1.$$

The first part is a simple consequence of compactness. The second part is an easy exercise. The third part is essentially the Hausdorff condition on the group.

We can now refine this definition to get our key player:

DEFINITION 2. A pro-p group is a profinite group in which every open subgroup has index equal to some power of a fixed prime p.

Before we explore examples of these groups, here is the more constructive definition of profinite and pro-p groups realised by stringing together finite groups using the inverse limit construction. Let us first recall how this is defined:

DEFINITION 3. (1) A directed set $(\Lambda, \leq)$ is a set with an ordering which satisfies the following criterion: if λ, $\mu \in \Lambda$ then there exists $\nu \in \Lambda$ such that $\nu \geq \lambda$ and $\nu \geq \mu$.
(2) An inverse system of groups $(G_\lambda)_{\lambda\in\Lambda}$ is a set of groups G_λ indexed by a directed set $(\Lambda, \leq)$ together with linking homomorphisms $\pi_{\lambda,\mu} \colon G_\lambda \to G_\mu$ between groups G_λ and G_μ whose indexes are comparable $\lambda \geq \mu$ with the following properties:
   (a) $\pi_{\lambda,\lambda} = \mathrm{Id}_{G_\lambda}$;
   (b) $\pi_{\lambda,\mu}\pi_{\mu,\nu} = \pi_{\lambda,\mu}$ for $\lambda \geq \mu \geq \nu$.
(3) An inverse limit of groups is a subgroup of the cartesian product of groups G_λ in an inverse system whose coordinates match up under the linking homomorphisms:

$$\varprojlim{}_{\lambda\in\Lambda} G_\lambda = \left\{ (g_\lambda) \in \prod_{\lambda\in\Lambda} G_\lambda \colon g_\lambda \pi_{\lambda\mu} = g_\mu \text{ for } \lambda \geq \mu \right\}.$$

We can now state the following:

PROPOSITION 2. (1) *G is a profinite group if and only if G is an inverse limit of finite groups.*
(2) *G is a pro-p group if and only if G is an inverse limit of finite p-groups.*

We dedicate some space now to presenting various examples of these groups.

2. Examples of pro-p Groups

EXAMPLE 1. An abstract group G with the discrete topology is a profinite group if and only if G is finite. It is a pro-p group if and only if it is a finite p-group.

EXAMPLE 2. The p-adic integers $\mathbb{Z}_p$ is a core building block in the theory of pro-p groups. As we shall see every pro-p group admits a natural action of $\mathbb{Z}_p$. We can view the p-adic integers as a pro-p group via its construction as an inverse limit of finite cyclic p -groups:

$$\mathbb{Z}_p = \varprojlim_{n\in\mathbb{N}} \mathbb{Z}/p^n\mathbb{Z}.$$

The p-adic integers plays the role of the infinite cyclic group in the theory of pro-p groups—it is in fact a pro-cyclic group. We can view this construction as taking the inverse limit of all finite p-quotients of the infinite cyclic group. As we shall see, this construction of the p-adic integers is a procedure we can carry out on any abstract group, not just the infinite cyclic group, to get a pro-p group. Before we explain this construction, we consider next some other examples of pro-p rings.

EXAMPLE 3. Another key character is the discrete valuation ring $\mathbb{F}_p[\![T]\!]$ of characteristic p. This is the inverse limit of quotients of the polynomial ring $\mathbb{F}_p[T]$ with respect to the subgroups $T^i\mathbb{F}_p[T]$. As we shall see it is rather a degenerate pro-p group by itself, but we get some very interesting pro-p groups when we plug it into entries of matrix group (see Example 8) and most interestingly when we consider the pro-p group arising from the automorphism group of the ring structure of $\mathbb{F}_p[\![T]\!]$ (see Example 10). The examples $\mathbb{Z}_p$ and $\mathbb{F}_p[\![T]\!]$ are both discrete valuation rings. Recall that a commutative local ring R with (unique) maximal ideal $\mathfrak{m}$ is a discrete valuation ring if and only if the maximal ideal $\mathfrak{m}$ is generated by a single element. But we can consider more general pro-p rings of Krull dimension greater than 1. A commutative pro-p ring R is defined to be a complete commutative Noetherian local ring $(R, \mathfrak{m})$ whose residue field $R/\mathfrak{m}$ is finite. For example the ring $\mathbb{Z}_p[\![T]\!]$. Such a ring is in fact the homomorphic image of $R_0[\![T_1, \ldots, T_n]\!]$ where R_0 is the prime ring of R, which in this topological setting means the closure of the ring generated by 1 (i.e. $\mathbb{Z}_p$ or $\mathbb{F}_p$).

We can also define a pro-p ring constructively as an inverse limit of finite local rings of size a power of p. These are equivalent definitions since in one direction a commutative pro-p ring is clearly a complete, Noetherian local ring and $R/\mathfrak{m}$ is finite. Conversely a Noetherian local ring $(R, \mathfrak{m})$ has the property that the intersection of all powers of the maximal ideal $\mathfrak{m}$ is trivial. So we can embed R into the inverse limit over quotients of the maximal ideal. Because it is complete it embeds surjectively. Because the residue field $R/\mathfrak{m}$ is finite, the quotients by powers of the maximal ideal are finite. Since $R/\mathfrak{m}$ is a field it must have order a power of some prime p. Hence these quotients rings are commutative pro-p rings.

Traditionally we have considered only examples of pro-p groups with an analytic structure built out of discrete valuation rings (see [**5, 30**]). But, as we shall see later, we introduce a definition of a class of pro-p groups which have an analytic structure over a more general pro-p ring of Krull dimension greater than 1. Lubotzky and Shalev introduced examples of these groups in their paper [**26**] and alluded to the possibility of a category of pro-p groups called R-analytic groups. We take up their challenge and propose a category into which their examples fit. But more of that later.

EXAMPLE 4. We alluded to a way to construct pro-p groups out of abstract which generalizes the construction of the p-adic integers as the inverse limit of p-quotients of the infinite cyclic group. Associated to every abstract group Γ there is a natural inverse system of finite groups just asking to be strung together into an inverse limit, namely the finite quotients Γ/N of the group Γ and the natural connecting maps between two quotients $\Gamma/N \to \Gamma/M$ whenever the two groups are comparable, i.e. $N \leq M$. Note that the indexing set for this inverse system is the set of normal subgroups of finite index ordered with respect to inverse inclusion. The inverse limit of this system gives rise to a profinite group called the *profinite completion* of the group and denoted $\widehat{\Gamma}$.

We needn't have taken all finite quotients but just some subsystem $(N_i)_{i\in I}$. Note that to be an inverse system we will require that for any two subgroups N_i and N_j there exists a third subgroup N_k contained in each of them. An example of such a subsystem is the set of normal subgroups of index a power of a fixed prime p. Since all the finite quotients are p-groups the corresponding inverse limit is then a pro-p group, called the *pro-p completion* of the group Γ and denoted by $\widehat{\Gamma}_p$.

We mentioned above that the p-adic integers are the pro-p completion of the infinite cyclic group. Note that the primary decomposition of an integer implies that the profinite completion of the integers is in fact the product of the p-adic integers over all primes p.

As another example, it is illustrative to think about what the profinite completion of the group $\mathrm{GL}_n(\mathbb{Z})$ looks like. You will soon get into some deep questions about something called the congruence subgroup problem and strong approximation. Lubotzky has championed the use of profinite groups as a tool for attacking this problem for more general matrix groups (see [**21, 22**]). In particular, there has been some recent activity in understanding systems of subgroups of an abstract group which have property τ, a generalization of Khazdan property T (see [**23**]).

A group is called residually-finite (respectively residually-finite-p) if the intersection of all the normal subgroups of finite index (respectively p-power index) is the identity element. Such groups have the important property that they embed into their profinite (respectively pro-p) completions. This is often the key to many applications of profinite groups to abstract groups (see [**9, 8**]).

Every profinite group and pro-p group arises out of this construction because they are the completions with respect to the corresponding system of open subgroups. For finitely generated pro-p groups we shall see that this is the same as the pro-p completion of the pro-p group, in other words pro-p groups have the property that every subgroup of finite index is in fact an open subgroup. (We say that a topological group is *finitely generated* if there exists a finite set of elements which generate a dense subset.) For finitely generated profinite groups this question is still open:

CONJECTURE 1. *Every subgroup of finite index in a finitely generated profinite group is open. Or equivalently: the profinite completion of a profinite group G is just G.*

What is more interesting is realising a finitely generated pro-p group as the pro-p completion of a finitely generated abstract group. For example, the fact that the quaternion group Q_{16} admits a four-dimensional representation over $\mathbb{Z}_2$, the 2-adic integers, but not over the rational integers $\mathbb{Z}$ means that the pro-2 group $\mathbb{Z}_2^{(4)} \times Q_{16}$, constructed as a semi-direct product using this action of Q_{16} on $\mathbb{Z}_2^{(4)}$, cannot be realised as the pro-p completion of a finitely generated abstract group.

But there is still alot of mileage to be had from checking what the pro-p completion of any group you might be handling looks like. As we shall see in Example 11 the pro-p completion of the group of Grigorchuk of intermediate word growth (see [**14**]) provided a recent counterexample to a conjecture characterizing pro-p groups whose lower central series have bounded sections, sometimes referred to as groups of finite width.

An important profinite group which arises out of this construction is given in the next example.

EXAMPLE 5. If $\Gamma(X)$ denotes the free abstract group on a finite set of generators X than its profinite completion is a free profinite group in the category of profinite groups. Similarly the pro-p completion gives a free pro-p group in the category of pro-p groups. We therefore have a sensible notion of presentations of profinite and pro-p groups: given a set of generators X and relations R, the profinite group given by the presentation $\langle X|R\rangle$ is then just the group $\widehat{\Gamma}(X)/N$ where N is the closure in the profinite topology of the normal closure of the subgroup generated by the relations R. For more details in this direction we refer the reader to [**20, 29**].

EXAMPLE 6. Profinite groups essentially appeared on the scene because they are how one defines the Galois group of an infinite field extension. Suppose L is an infinite Galois field extension of the field K. We can realise L as a union of finite Galois field extensions L_i of K

$$L = \bigcup_{i \in I} L_i.$$

If $L_i \geq L_j$ then there is a natural restriction map between the corresponding finite Galois groups $\mathrm{Gal}(L_i{:}K) \to \mathrm{Gal}(L_j{:}K)$. Together with these maps this makes $\mathrm{Gal}(L_i{:}K)_{i\in I}$ into an inverse system whose inverse limit we call the Galois group $\mathrm{Gal}(L{:}K)$ of the field extension $(L{:}K)$:

$$\mathrm{Gal}(L{:}K) = \varprojlim_{i\in I} \mathrm{Gal}(L_i{:}K).$$

What groups can occur as Galois groups of extensions of a given field K, especially for $K = \mathbb{Q}$, (the so-called inverse Galois problem) has become big business since the classification of finite simple groups. Every pro-p group can be realised as a Galois group of some field extension L/K. If we fix for example the field K to be a local number field of degree n over $\mathbb{Q}_p$ then Shafarevich proved in [**31**] that, provided K does not contain p-th roots of unity then every $n+1$-generated pro-p group can be realised as the Galois group of some field extension L/K. Another way to say this is that the maximal p-extension L/K has the free pro-p group on $n+1$ generators as Galois group (a p-extension is an extension whose Galois group is a p-group). If however K contains p-th roots of unity, then the maximal p-extension can be defined by a pro-p presentation with a finite set of generators with one relation of a special type. These groups are called Demushkin groups after Demushkin who made the first in-roads into classifying the shape of these Galois groups. Labute [**18**] was responsible for completing the classification.

Another direction which has seen some stimulating interplay between field theory and pro-p groups is the class field tower problem. Here we build up an extension of a global number field, extending the field of rational numbers by a tower of finite field extensions L_i of a special type: namely, L_i should be the maximal unramified abelian extension of the previous field L_{i-1}. Class field theory is concerned with determining all the abelian extensions of a given field, i.e. those extensions with abelian Galois groups, and tells us that such a maximal extension does exist, called the Hilbert Class Field. The union of these field extensions is the maximal unramified extension L of K. The class tower problem asks whether this tower of field extensions stabilizes after finitely many steps.

This question, posed by Furtwangler in 1926, was answered by Golod and Shafarevich [**12**] using the theory of pro-p groups. They showed that the field

$$K = \mathbb{Q}\big(\sqrt{-3.5.7.11.13.17.19}\big)$$

has an infinite class tower above it. To do this they proved that a presentation for the maximal p-quotient $G(p)$ of the Galois group G of L/K could not be the presentation of a finite p-group. Presentations of finite p-groups have the property that given $h_1(G)$ generators for the group you need a presentation with at least $h_2(G) > h_1(G)^2/4$ relations. This is called the Golod-Shafarevich inequality. It implies that the difference $\sigma(G) = h_2(G) - h_1(G)$ tends to infinity as $h_1(G)$ tends to infinity. However if L has no cyclic unramified extensions of degree p then $\sigma(G) \leq r_1 + r_2$ where r_1 (resp. r_2) denotes the number of real (resp. pairs of complex) conjugates of K. In this case $r_1 + r_2 = 1$. But $G(p)$ has 7 generators, the number of primes we took in defining K. Hence if $G(p)$ was a finite p-group, satisfying the Golod-Shafarevich inequality, $h_2(G) > 12$ whilst $h_1(G) = 7$. This would contradict $\sigma(G) = h_2(G) - h_1(G) \leq r_1 + r_2 = 1$.

It has subsequently been shown that the argument that shows that a finite p-group satisfies the Golod-Shafarevich inequality, applies equally to a pro-p group which has a p-adic analytic structure (see [**17, 21**]) and more recently a pro-p group with an analytic structure over any pro-p ring (see [**26**]). Hence the maximal p-quotient of the Galois group of the infinite class field tower of K cannot have an analytic structure since the above argument only depends on the Galois group $G(p)$ not satisfying the Golod-Shafarevich inequality.

What can we say about this pro-p group and others which are the Galois groups of maximal unramified p-extensions which are infinite? One form of a conjecture due to Fontaine and Mazur asserts that actually this group will never have an infinite p-adic analytic quotient (see [**11**]). Boston has proved this interpretation of the Fontaine-Mazur conjecture for a large class of fields (see [**2, 3**]). In [**3**] he lists various conditions that such pro-p groups must satisfy and challenges group theorists to classify groups with these properties.

Example 7. The group $\mathrm{GL}_n(\mathbb{Z}_p)$ is a fundamental example. It is a profinite group but by looking at the exact sequences

$$1 \to \mathrm{GL}_n^i(\mathbb{Z}_p) \to \mathrm{GL}_n(\mathbb{Z}_p) \to \mathrm{GL}_n(\mathbb{Z}_p/p^i\mathbb{Z}_p) \to 1$$

we see that it is the extension of a pro-p group $\mathrm{GL}_n^1(\mathbb{Z}_p)$, called the first principal congruence subgroup, by the finite group $\mathrm{GL}_n(\mathbb{F}_p)$. The matrix representation gives us a natural coordinate system defined on the group. With respect to this coordinate system, group multiplication is given by polynomial functions. However taking the inverse of a matrix will require a power series in the entries. This is the key example of an analytic group over $\mathbb{Z}_p$. We shall make a more formal definition of an analytic group later but essentially it is a group like $\mathrm{GL}_n(\mathbb{Z}_p)$ on which we can put a coordinate system with respect to which group operations are given by formal power series in the coordinates. In fact profinite groups analytic over $\mathbb{Z}_p$ can always be realised as closed subgroups of our example $\mathrm{GL}_n(\mathbb{Z}_p)$.

Example 8. We can realise other profinite groups by considering GL_n over other pro-p rings, for example $\mathrm{GL}_n\big(\mathbb{F}_p[\![T]\!]\big)$ or $\mathrm{GL}_n\big(\mathbb{Z}_p[\![T]\!]\big)$. More generally if we set R to be a commutative local pro-p ring with unique maximal ideal $\mathfrak{m}$, then

$\mathrm{GL}_n(R)$ is a profinite group. Again we have an exact sequence

$$1 \to \mathrm{GL}_n^1(R) \to \mathrm{GL}_n(R) \to \mathrm{GL}_n(R/\mathfrak{m}) \to 1$$

and $\mathrm{GL}_n^1(R)$ is a pro-p group. These examples look like they should be called analytic over the ring R since we have a coordinate system and power series defining the group operations. In the case of the ring $\mathbb{F}_p[\![T]\!]$ we have a well defined category of groups analytic over $\mathbb{F}_p[\![T]\!]$ since this ring is a discrete valuation ring. But more generally, no such category has been defined. We will look in the later part of this paper to make such a definition which will include $\mathrm{GL}_n(R)$ as an example.

If we take instead of GL_n, some other Chevalley group scheme with entries in R then we can realise more examples of what we might hope to call R-analytic groups. More generally, suppose that R is an integral domain with field of fractions K. Let G be a linear algebraic group defined over the field of fractions K and take some faithful K-rational representation $\phi\colon G \to \mathrm{GL}_n$. Then $G(R) = \phi^{-1}\big(\phi G(K) \cap \mathrm{GL}_n(R)\big)$ is a profinite group which deserves to be called analytic over the ring R. It contains a pro-p group of finite index which is the kernel of the map $G(R) \to \mathrm{GL}_n(R/\mathfrak{m})$.

Example 9. We can also realise some of the examples of the previous section as subgroups of loop groups. Classically loop groups LG are defined as the group of parameterized loops in a group G or in other words the group of maps from the circle S^1 into G. The concept arose in the realm of quantum field theory where the group G is a Lie group over $\mathbb{R}$ or $\mathbb{C}$ (see [**27**]). But instead we could take G to be a subgroup of $\mathrm{GL}_n(R)$ where R is some pro-p ring like $\mathbb{Z}_p$ and take the loop group to be $\mathrm{Map}(R; G)$, the set of continuous maps from R into G. Each loop γ can be interpreted as a matrix function $z \mapsto \big(\gamma_{ij}(z)\big)$. We can then consider subgroups of this loop group, for example where the functions $\gamma_{ij}(z)$ are defined by strictly analytic functions $\gamma_{ij}(z) = \sum_{k=0}^{\infty} a_{ijk} z^k$ where $a_{ijk} \in R$. Such subgroup look either like infinite dimensional Lie groups over the base ring R or alternatively look like Lie groups over the ring of Laurent series with coefficients from R. We call this *the positive part of the loop group over G*. Such groups also deserve to be called analytic groups over the power series ring $R[\![T]\!]$.

If the Lie group we started with is a Chevalley group with points in R then the positive part of the loop group will be the Chevalley group with points in $R[\![T]\!]$.

If G is a Lie group over R then it has a Lie algebra $\mathfrak{g}$ defined over the field of fractions of R. The Lie algebra of the associated loop group LG is the set of loops from R to the Lie algebra $\mathfrak{g}$. These turn out to be examples of Kac-Moody algebras over the field K. It is actually a particular type of Kac-Moody algebra called an affine Lie algebra (see [**16**]).

Although continuous maps from $\mathbb{F}_p$ to itself are rather limited, it is natural to define a group G with filtration G_i whose graded Lie algebra $L(G) = \bigoplus_{i \in \mathbb{N}} G_i/G_{i+1}$ is the positive part of an affine Lie algebra or loop algebra over $\mathbb{F}_p[\![T]\!]$ to be the positive part of the loop group from $\mathbb{F}_p$. For example, if $G = \mathrm{SL}_n^1\big(\mathbb{F}_p[\![T]\!]\big)$ with filtration $G_i = \mathrm{SL}_n^i\big(\mathbb{F}_p[\![T]\!]\big)$ then as we shall see in Section 4 the graded Lie algebra $L(G) = \mathfrak{sl}_n(\mathbb{F}_p) \otimes \mathbb{F}_p[\![T]\!]$, is the positive part of the loop algebra on $\mathfrak{sl}_n(\mathbb{F}_p)$. Hence $\mathrm{SL}_n^1\big(\mathbb{F}_p[\![T]\!]\big)$ is sometimes referred to as the positive part of a loop group.

Example 10 (The Nottingham group). The example in this section is a Galois group but one which has only recently received any serious attention by group theorists. It turns out to be a very stimulating group with some unexpected properties.

It is a subgroup of the group $\mathrm{Aut}(\mathbb{F}_p[\![T]\!])$ of ring automorphisms of the ring $\mathbb{F}_p[\![T]\!]$. This is the same in fact as the Galois group $\mathrm{Gal}(\mathbb{F}_p((T)) : \mathbb{F}_p)$ where $\mathbb{F}_p((T))$ is the field of fractions of $\mathbb{F}_p[\![T]\!]$. Any automorphism $f \in \mathrm{Aut}(\mathbb{F}_p[\![T]\!])$ is determined by its action on T. We can use this to identify $\mathrm{Aut}(\mathbb{F}_p[\![T]\!])$ with the set of power series $\{\sum_{i=1}^{\infty} a_i T^i : a_i \in \mathbb{F}_p, a_1 \neq 0\}$ where the group operation corresponds to substitution of power series. This group is not a pro-p group but it contains a pro-p group of index $p-1$, called the Nottingham group after its pioneer Dave Johnson of the University of Nottingham:

$$\begin{aligned}\mathrm{Nott}(\mathbb{F}_p) &= \{f \in \mathrm{Aut}(\mathbb{F}_p[\![T]\!]) : f \text{ acts trivially on } T\mathbb{F}_p[\![T]\!]/T^2\mathbb{F}_p[\![T]\!]\} \\ &= \left\{T + \sum_{i=2}^{\infty} a_i T^i : a_i \in \mathbb{F}_p\right\}_{\text{substitution}}.\end{aligned}$$

We can in fact generalize this example to consider automorphisms of $R[\![T]\!]$ for any pro-p ring R and the corresponding Nottingham group $\mathrm{Nott}(R)$.

Al Weiss and Charles Leedham-Green proved that this group has the following property:

THEOREM 1. *The group* $\mathrm{Nott}(\mathbb{F}_p)$ *contains every finite p-group as a subgroup.*

Their proof is based on work of Witt's from the 1930's. Witt showed how to realise any finite p-group H as a Galois group of a finite extension K of $\mathbb{F}_p((T))$. Hence $H \leq \mathrm{Aut}(K)$. But then one observes that K itself is isomorphic to $\mathbb{F}_p((T'))$ being a finite totally ramified extension of $\mathbb{F}_p((T))$. Hence

$$H \leq \mathrm{Aut}(K) = \mathrm{Aut}(\mathbb{F}_p((T'))).$$

Since H is a p-group it follows that H is in fact a subgroup of

$$\mathrm{Nott}(\mathbb{F}_p) \leq \mathrm{Aut}(\mathbb{F}_p((T'))).$$

Rachel Camina (see [**6**]) subsequently extended Weiss and Leedham-Green's result to show that you can in fact string together these finite p-groups sitting inside the Nottingham group to realise any finitely generated pro-p group as a closed subgroup of the Nottingham group:

THEOREM 2. *The group* $\mathrm{Nott}(\mathbb{F}_p)$ *contains every finitely generated pro-p group as a closed subgroup.*

As a corollary of these results we get the following important fact:

COROLLARY 1. *The group* $\mathrm{Nott}(\mathbb{F}_p)$ *is not a linear group over any field.*

We shall also see later that there is a natural graded Lie algebra associated with $\mathrm{Nott}(\mathbb{F}_p)$ which looks like the positive part of the Loop algebra of the first Witt Lie algebra W_1, one of the non-classical simple Lie algebras in characteristic p, the so-called Lie algebras of Cartan type.

By considering automorphisms of $R = \mathbb{F}_p[\![T_1, \dots, T_n]\!]$ which fix a form s on $R^{(n)}$ (for example $s = dT_1 \wedge \cdots \wedge dT_n$), Shalev and Leedham-Green hint at analogues of the Nottingham group whose graded Lie algebras realise the other Lie algebras of Cartan type.

EXAMPLE 11 (Fractal groups). The last examples that we present in this section are old groups but whose pro-p completions are providing an interesting new landscape in the theory of pro-p groups. In particular they throw some interesting light on a project to classify pro-p groups of *bounded lower-central width* which we shall define later.

These groups are defined as special subgroups of the automorphism group of a p-regular tree. They arose in a number of contexts, for instance examples of Aleshin [**1**], Sushchansky [**34**], Grigorchuk [**13**] and Gupta and Sidki [**15**] of infinite Burnside groups (finitely generated periodic groups) and examples of Grigorchuk [**14**] of groups with intermediate word growth.

Grigorchuk's example of a group of intermediate growth is defined by four generators a, b, c, d of the 2-regular tree which can be described as follows:

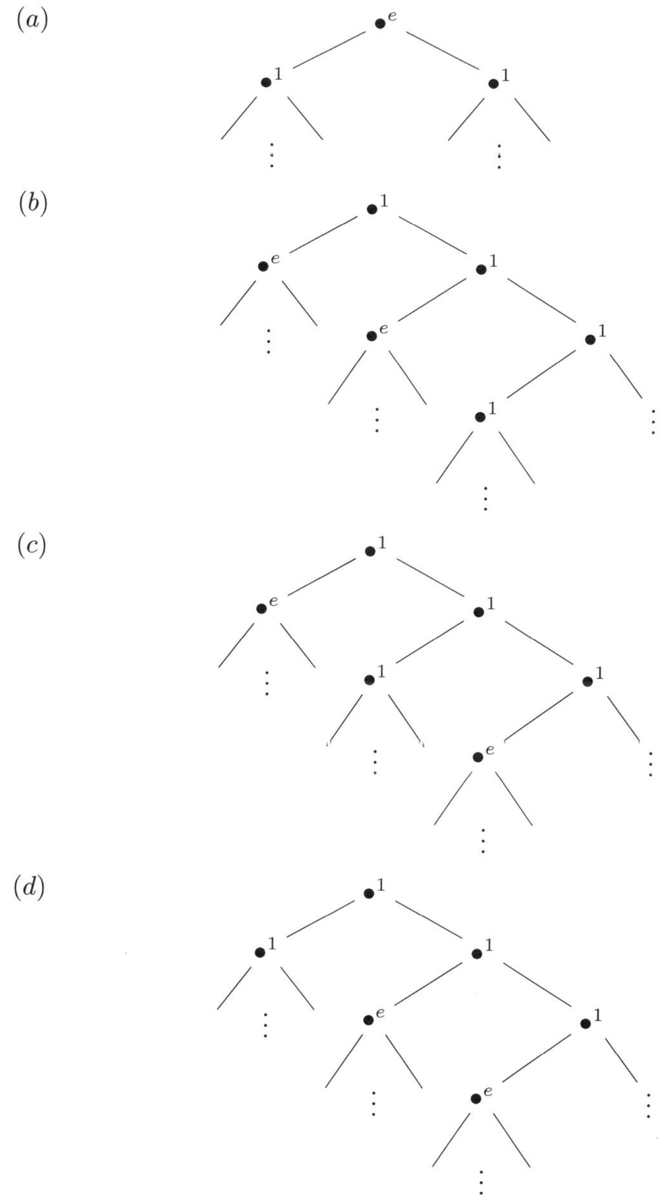

The convention used is that if e appears on a vertex then the automorphism swaps the edges below the vertex whilst 1 on a vertex leaves the edges the same. In generators b, c, d the pattern repeats itself after three vertices.

This example fits into a construction which has received general treatment by Rozhkov [**28**] which he calls groups of Aleshin type after the author of the first paper to define such examples [**1**]. Grigorchuk has suggested the name *Fractal Groups* due to the reproductive character of generators like a, b, c, d.

Rozhkov has recently calculated the lower central series of Grigorchuk's group $\langle a, b, c, d\rangle$. It turns out that the sections have bounded order. Its pro-p completion will have the same property and is an example of a group we shall later define as a group of finite lower-central width. A previous guess at the classification of such groups of finite lower-central width included the examples in 7, 8, 9 and 10 but had not included such examples of subgroups of tree automorphisms. It seems at present that the Gupta-Sidki group will not have finite lower-central width although this has not yet been verified.

The pro-p completion of Grigorchuk's group also shares the property with the Nottingham group that it contains every finite p-group and every finitely generated pro-p group as a closed subgroup. This is because of the existence of subgroups isomorphic to wreath products of the cyclic group of order p.

It seems likely that these pro-p fractal groups are going to play an important role in the theory of pro-p groups.

3. Pro-p Groups: The Basics

We present here some of the basic properties of pro-p groups which will be useful to those coming to the theory for the first time. Proofs can be found in the book [**8**].

DEFINITION 4. (1) A topological group is *finitely generated* if it is the closure of a finitely generated subgroup i.e. $G = \overline{\langle X\rangle}$ for some finite set X.
(2) We define $d(G) = \inf\{\text{card}(X) : G = \overline{\langle X\rangle}\}$.
(3) The rank of a group G, denoted by $r(G)$, is defined to be

$$\sup\{d(H) \colon H \text{ is a finitely generated subgroup of } G\}.$$

(4) The Frattini subgroup is defined to be

$$\Phi(G) = \bigcap\{M \colon M \text{ maximal proper open subgroup of } G\}.$$

When we talk about finite generation for a profinite group we shall always mean in the topological sense of Definition 4(1). As for finite groups, the Frattini subgroup is the set of non-generators of a profinite group.

PROPOSITION 3. *Let G be a profinite group. For a subset X of G the following are equivalent*:
(i) *X generates G*;
(ii) *$X \cup \Phi(G)$ generates G; and*
(iii) *$X\Phi(G)/\Phi(G)$ generates $G/\Phi(G)$.*

For a pro-p group we have an explicit description of the Frattini subgroup:

PROPOSITION 4. *Let G be a pro-p group*
(1) $\Phi(G) = \overline{G^p[G, G]}$.

(2) *G is finitely generated if and only if $\Phi(G)$ is open in G.*
(3) $d(G) = \dim_{\mathbb{F}_p}\big(G/\Phi(G)\big)$.

Note that we do not have such a description of the Frattini subgroup in general for a profinite group and that G can be finitely generated even though $\Phi(G)$ is trivial, for example the profinite completion of the integers $G = \widehat{\mathbb{Z}}$.

We can extend the definition of the Frattini subgroup of a pro-p group to define a key filtration associated to a pro-p group:

DEFINITION 5. Let G be a pro-p group. We define the *lower p-series* of G denoted $\{P_i(G)\}_{i=1}^{\infty}$ as follows:

$$P_1(G) = G$$
$$P_{i+1}(G) = \overline{P_i(G)^p[P_i(G), G]}.$$

In words, it is the fastest descending series of closed normal subgroups with each factor central and of exponent dividing p.

PROPOSITION 5. (1) *$P_2(G) = \Phi(G)$ and $P_{i+1}(G) \geq \Phi\big(P_i(G)\big)$.*
(2) $\big[P_i(G), P_j(G)\big] \leq P_{i+j}(G)$.

It is a crucial series in understanding the structure of a finitely generated pro-p group. In particular it defines the topology of such a group, i.e. it is a base of neighbourhoods as we record in the following:

PROPOSITION 6. *Suppose that G is a finitely generated pro-p group.*

(1) *$P_i(G)$ is open and $\{P_i(G) \colon i \geq 1\}$ is a base of neighbourhoods of the identity in G.*
(2) *The derived group $[G, G]$ is a closed subgroup.*
(3) *As a consequence of (2) we have that $P_{i+1}(G) = P_i(G)^p\big[P_i(G), G\big]$.*
(4) *This in turn implies that every subgroup of finite index is in fact open and hence the topology of a finitely generated pro-p group is determined by its group structure.*

It is conjectured that part (4) also holds true for finitely generated profinite groups but as yet this is still open (cf. Conjecture 1).

1. Examples: lower p-series.

EXAMPLE 12. If $G = \mathbb{Z}_p$ then G is a one generated pro-p group whose lower p-series is $p^i\mathbb{Z}_p$.

EXAMPLE 13. If $G = \mathbb{F}_p[\![T]\!]$ then G is not finitely generated and in fact every term of the lower p-series is trivial.

EXAMPLE 14. Consider instead of the additive group of the ring $\mathbb{F}_p[\![T]\!]$, the multiplicative structure. To get a pro-p group we need to take a subgroup of the units given by power series with 1 as constant coefficients:

$$G = \left\{1 + \sum_{i=1}^{\infty} \alpha_i T^i\right\}.$$

Then $\Phi(G) = G^p = \{1 + \sum_{i=1}^{\infty} \alpha_i T^{p^i}\}$. Since the Frattini subgroup has infinite index, G is infinitely generated. The lower p-series is given by $G_{n+1} = G^{p^n} = \{1 + \sum_{i=1}^{\infty} \alpha_i T^{p^{n}i}\}$.

EXAMPLE 15. The multiplicative group $\mathbb{Z}_p^*$ has a subgroup of finite index which is isomorphic to the additive structure under the logarithm map, hence we get the same picture as Example 12, in particular $\mathbb{Z}_p^*$ is finitely generated.

More generally we can consider the multiplicative group of units U_K of a local field K. Let ϑ_K denote the ring of integers of K and $\mathfrak{m}_K$ the maximal ideal in ϑ_K. For each $n \in \mathbb{N}$ define $U_n = 1 + \mathfrak{m}_K^n$. Then U_1 is a pro-p group. How does the filtration U_n compare with the lower p-series? Since U_1 is abelian this depends on how the p-power map operates which is detailed in the following Lemma. Let e_K denote the absolute ramification index of K.

LEMMA 1. *Let* $n \in \mathbb{N}$.
(i) *If* $n \leq e_K/(p-1)$ *then* $U_n^p \leq U_{np}$.
(ii) *If* $n > e_K/(p-1)$ *then* $x \mapsto x^p$ *defines an isomorphism from* U_n *onto* U_{n+e_K}.

EXAMPLE 16. We can generalize Examples 14 and 15 to the general linear groups over $\mathbb{Z}_p$ and $\mathbb{F}_p[\![T]\!]$. We begin with the case of $\mathrm{GL}_n(\mathbb{Z}_p)$. To get a pro-$p$ group, we consider the congruence subgroups $G_i = \mathrm{GL}_n^i(\mathbb{Z}_p)$ defined in Example 7, by the exact sequence

$$1 \to \mathrm{GL}_n^i(\mathbb{Z}_p) \to \mathrm{GL}_n(\mathbb{Z}_p) \to \mathrm{GL}_n(\mathbb{Z}_p/p^i\mathbb{Z}_p) \to 1.$$

We calculate the lower p-series $P_i = P_i(G_1)$ of G_1. We have a very easy description of the congruence kernels:

$$G_i = \mathrm{GL}_n^i(\mathbb{Z}_p) = 1 + p^i\,\mathrm{M}_n(\mathbb{Z}_p).$$

Using the identities $(1+X)(1+Y) = 1 + X + Y + XY$ and $(1+X)^p = 1 + X^p$ (mod p) we can show that G_i/G_{i+1} is an elementary abelian p-group and central in G/G_{i+1}. Hence $P_{i+1} \leq G_{i+1}$ since the series P_{i+1} is maximal with respect to these properties. In fact we can prove that we have an equality between the lower p-series and the congruence kernels. The crucial step is to show that every element of G_{i+1} can be written as the p-th power of an element in G_i, i.e. $G_i^p \geq G_{i+1}$. This requires an application of Hensel's lemma to solve for $x \in \mathrm{M}_n(\mathbb{Z}_p)$ the following matrix equation:

$$(1 + p^{i+1}a) = (1 + p^i x)^p$$

where $a \in \mathrm{M}_n(\mathbb{Z}_p)$. We leave this application of Hensel's lemma as an exercise but our first approximation to get Hensel's lemma going is of course to take $x = a$.

Suppose by induction we have already established that $G_i = P_i$ then

$$G_{i+1} \geq P_{i+1} \geq P_i^p \geq G_i^p \geq G_{i+1}.$$

Thus $G_n = P_n$ for all n. In fact we have shown more which reveals a crucial character of pro-p groups which are analytic over $\mathbb{Z}_p$, namely that the $(n+1)$-th term of the lower p-series is precisely the set of p^n-powers of the group G: $P_{n+1} = G^{p^n}$.

The application of Hensel's lemma here is rather intriguing. Is it possible to formulate a form of Hensel's lemma in the language of pro-p groups? For example if we have a group equation which we are trying to solve and we have a solution modulo some term of the lower p-series, when can we know that this lifts to a solution in the pro-p group? This might have some interesting ramifications for the model theory of pro-p groups as Hensel's lemma is the key to understanding the model theory of the p-adic integers.

Generalizing the last paragraph of Example 15, Weigel [**35**] has identified the lower p-series for Chevalley groups over a number field in terms of the ramification index of the field.

EXAMPLE 17. If we look at the congruence subgroups of $\mathrm{GL}_n(\mathbb{F}_p[\![T]\!])$ then since the example $\mathbb{F}_p[\![T]\!]^*$ occurs as a quotient of $\mathrm{GL}_n^1(\mathbb{F}_p[\![T]\!])$ on taking the determinant map, we see that $\mathrm{GL}_n^1(\mathbb{F}_p[\![T]\!])$ is also an infinitely generated pro-p group.

However things change for the better if we consider instead the subgroup $\mathrm{SL}_n(\mathbb{F}_p[\![T]\!])$ and its congruence subgroups $G_i = \mathrm{SL}_n^i(\mathbb{F}_p[\![T]\!]) \leq 1 + T^i\,\mathrm{M}_n(\mathbb{F}_p[\![T]\!])$. Since we are working in characteristic p the p-power map sends elements much deeper in the group than we experienced in Example 16 since $(1 + T^iX)^p = 1 + T^{ip}X^p$. Hence $G_i^p \leq G_{pi}$.

However the lower p-series still turns out to coincide with the congruence subgroups. This is due to the way the congruence subgroups commutate: $[G_i, G_j] = G_{i+j}$. We indicate a proof of this fact. The matrix identity $[1+T^iX, 1+T^jY] = 1 + T^{i+j}(XY - YX) \pmod{T^{i+j+1}}$ proves that $[G_i, G_j] \leq G_{i+j}$. The converse follows from the key fact that the Lie algebra $\mathfrak{sl}_n(\mathbb{F}_p)$ is perfect, i.e. for all $Z \in \mathfrak{sl}_n(\mathbb{F}_p)$ there exist $X_1, \ldots, X_l, Y_1, \ldots, Y_l \in \mathfrak{sl}_n(\mathbb{F}_p)$ such that $Z = (X_1, Y_1) + \cdots + (X_l, Y_l)$ where $(\ ,\ )$ is the Lie bracket in $\mathfrak{sl}_n(\mathbb{F}_p)$. Using our matrix identity again we can then translate this into being able to express every element of G_{i+j} as an element of $[G_i, G_j]$ mod G_{i+j+1}: For every $1 + T^{i+j}Z \in G_{i+j}$ there exist $1 + T^iX_1, \ldots, 1 + T^iX_l \in G_i$ and $1 + T^jY_1, \ldots, 1 + T^jY_l \in G_j$ such that

$$\begin{aligned} 1 + T^{i+j}Z &= 1 + T^{i+j}\big((X_1, Y_1) + \cdots + (X_l, Y_l)\big) \mod G_{i+j+1} \\ &= [1 + T^iX_1, 1 + T^jY_1] \ldots [1 + T^iX_l, 1 + T^jY_l] \mod G_{i+j+1}. \end{aligned}$$

A simple induction argument will get us for all k that $G_{i+j} = [G_i, G_j] \pmod{G_{i+j+k}}$ and since the congruence kernels are a base of neighbourhoods for the identity, this implies in turn that $G_{i+j} = [G_i, G_j]$.

We have thus established the coincidence of the congruence subgroups, the lower p-series and the lower central series:

$$P_n(G) = G_n = \gamma_n(G).$$

Hence in contrast to $\mathrm{GL}_n^1(\mathbb{F}_p[\![T]\!])$, the group $\mathrm{SL}_n^1(\mathbb{F}_p[\![T]\!])$ is finitely generated. The crucial step was the fact that the Lie algebra $\mathfrak{sl}_n(\mathbb{F}_p)$ is perfect. Lubotzky and Shalev [**26**] exploit this observation to good effect to identify a sensible class of groups with an analytic structure over $\mathbb{F}_p[\![T]\!]$ to study. We shall come to their definition later.

Notice though that since the p-th power map pushes G_n inside G_{pn}, if we calculate the Frattini subgroup of G_n then $\Phi(G_n) = G_{2n}$. Hence $d(G_n) \to \infty$ as $n \to \infty$, i.e. G has infinite rank.

EXAMPLE 18. The Nottingham group $G = \mathrm{Nott}(\mathbb{F}_p)$ also has a natural set of "congruence subgroups" G_n defined by the following exact sequence:

$$1 \to G_n \to G \to \mathrm{Aut}\big(\mathbb{F}_p[\![T]\!]/T^{n+1}\mathbb{F}_p[\![T]\!]\big) \to 1.$$

$\{G_n\}$ is a base of neighbourhoods of the identity and $|G_n \colon G_{n+1}| = p$. As a corollary, this gives us that the Nottingham group is in fact a pro-p group. There is a natural choice of generators for the layers G_n/G_{n+1}:

DEFINITION 6. Let $e_n \in G_n$ denote the automorphism have the following action on T:

$$(T)e_n = T(1+T^n).$$

PROPOSITION 7. (1) $\langle e_n \rangle G_{n+1} = G_n$.

(2) $G_n = \overline{\langle e_n, e_{n+1}, \dots \rangle}$.

(3) $[G_m, G_n] = \begin{cases} G_{n+m} & \textit{if } m \not\equiv n \mod p \\ G_{n+m+1} & \textit{if } m \equiv n \mod p \end{cases}.$

(4) $G_n^p \leq G_{np}$.

We can use the information in this Proposition to identify the lower-p series $P_i(G)$ and central series $\gamma_i(G)$ of G.

PROPOSITION 8. (1) *The lower central series of* $G = \mathrm{Nott}(\mathbb{F}_p)$ *looks like*

γ_1	G_1 | G_2	p^2	
γ_2	G_3	p	$\left.\right\}$ $p-2$
$\vdots$	$\vdots$	$\vdots$	
γ_{p-1}	G_p	p	
γ_p	G_{p+1} | G_{p+2}	p^2	
γ_{p+1}	G_{p+3}	p	
$\vdots$	$\vdots$	$\vdots$	

(2) *The lower p-series coincides with the lower central series*: $P_i(G) = \gamma_i(G)$.

(3) $\Phi(G_m) = G_{2m+1}$ *so G has infinite rank and is therefore not a p-adic analytic group as we shall see.*

(4) *G is finitely generated*: $G = \overline{\langle e_1, e_2 \rangle}$.

Although we in fact arrived at the same conclusion for the calculation of the lower p-series in both the example $\mathrm{GL}_n^1(\mathbb{Z}_p)$ and $\mathrm{SL}_n^1(\mathbb{F}_p[\![T]\!])$, there is a very different philosophy involved in each case. Over $\mathbb{Z}_p$ the p-th powers G^{p^n} play a dominant role in calculating the lower p-series; whilst this contrasts with the group over $\mathbb{F}_p[\![T]\!]$ where p-th powers disappeared too fast and the lower p-series is dominated by the commutator subgroups. We also saw that this property of the p-th power forces the group to be of infinite rank. We will see this is a prevalent philosophy.

Where the p-th powers dominate we in fact have a very satisfying theory as we shall now explain. It can be summed up as saying that in a p-adic analytic pro-p group, p-th powers dominate and conversely, a pro-p group in which p-th powers dominate actually can be seen to have an underlying p-adic structure.

The dominance of the p-th powers is captured in the next definition of a group "full of powers":

DEFINITION 7. Let G be a pro-p group. Then G is called *powerful* if

(a) $p > 2$ and $\overline{G^p} \geq [G, G]$ (i.e. $G/\overline{G^p}$ is abelian); or

(b) $p = 2$ and $\overline{G^4} \geq [G, G]$ (i.e. $G/\overline{G^4}$ is abelian).

Note that for $p = 2$, $G/\overline{G^2}$ is always abelian.

The lower p-series of a powerful pro-p group has a particularly nice description:

PROPOSITION 9. *Let G be a finitely generated powerful pro-p group. Put* $G_i = P_i(G)$. *Then*

(1) $G_i = G^{p^{i-1}} = \{x^{p^{i-1}} : x \in G\}$;
(2) $x \mapsto x^{p^k}$ *induces an epimorphism*

$$G_i/G_{i+1} \twoheadrightarrow G_{i+k}/G_{i+k+1}.$$

Note that a priori $G^p \neq \{x^p | x \in G\}$ and $(G^p)^p \neq G^{p^2}$.

We can draw a picture of a powerful pro-p group in layers, where the i-th layer is given by G_i/G_{i+1}, which is isomorphic to a finite-dimensional vector space over $\mathbb{F}_p$.

$G/G_2 \cong \mathbb{F}_p^{d(G)}$	G/G_2	G	powerful
$G_2/G_3 \cong \mathbb{F}_p^{d_2}$	G_2/G_3	G_2	
$G_3/G_4 \cong \mathbb{F}_p^{d_3}$	G_3/G_4	G_3	
$G_4/G_5 \cong \mathbb{F}_p^{d}$	G_4/G_5	G_4	uniform
$G_5/G_6 \cong \mathbb{F}_p^{d}$	G_5/G_6		
	$\vdots$		
$G_i/G_{i+1} \cong \mathbb{F}_p^{d}$	G_i/G_{i+1}	$G_i \ \downarrow g^p \ G_{i+1}$	
	$\vdots$		

By Proposition 9 the p-th power map takes each layer onto the subsequent layer. Hence the dimensions of these layers must decrease until at some point they stabilize. We call the subgroup at which this process stabilizes a *uniform* pro-p group. It is the group that Lazard originally called p-saturable in his ground-breaking work on p-adic analytic groups (see [**19**]).

DEFINITION 8. A pro-p group G is called *uniform* if G is a finitely generated powerful pro-p group and for all $i \geq 1$

$$|G\colon P_2(G)| = |P_i(G)\colon P_{i+1}(G)| = p^d.$$

We call $d = \dim(G)$ the *dimension* of the group G.

Proposition 9 combined with our picture established the following:

THEOREM 3. *Let G be a finitely generated powerful pro-p group, then $P_k(G)$ is uniform for k sufficiently large.*

2. Example: a uniform pro-p group. Our analysis in Example 16 of the congruence subgroup $\mathrm{GL}_n^1(\mathbb{Z}_p)$ suffices to prove that, for $p > 2$, $\mathrm{GL}_n^1(\mathbb{Z}_p)$ is a uniform pro-p group of dimension d^2. For $p = 2$ we must take the subgroup $\mathrm{GL}_n^2(\mathbb{Z}_2)$. Note that although the lower p-series of $\mathrm{GL}_n^1(\mathbb{F}_p[\![T]\!])$ looks as uniform as that of $\mathrm{GL}_n^1(\mathbb{Z}_p)$, the group $\mathrm{GL}_n^1(\mathbb{F}_p[\![T]\!])$ is not powerful.

Lazard saw the intimate connection between uniform pro-p groups and p -adic analytic groups. We shall come to define a p-adic analytic group in the next section, but without much analysis we can build up much of the Lie theory associated with a uniform group.

Lazard showed how to build an intrinsic p-adic manifold structure or coordinate system directly onto a uniform pro-p group by exploiting a natural action of the p-adic integers on any pro-p group. This action depends on the following:

LEMMA 2. *Let G be a pro-p group and $g \in G$. Let a_i and $b_i \in \mathbb{Z}$, $i \in \mathbb{N}$ such that*

$$\lim a_i = \lim b_i = \lambda \in \mathbb{Z}_p.$$

Then (g^{a_i}) and (g^{b_i}) converge in G to the same limit.

DEFINITION 9. $g^\lambda = \lim g^{a_i}$ where $\lambda = \lim a_i \in \mathbb{Z}_p$.

THEOREM 4. *Let G be a uniform pro-p group of dimension d topologically generated by elements $a_1, \ldots, a_d$. Then if $x \in G$ there exist unique $\lambda_1, \ldots, \lambda_d \in \mathbb{Z}_p$ such that*

$$x = a_1^{\lambda_1} \ldots a_d^{\lambda_d}.$$

The map $\phi\colon G \to \mathbb{Z}_p^d$ defined by $x \mapsto (\lambda_1, \ldots, \lambda_d)$ is a homeomorphism.

We can prove this by successive approximation with respect to the lower p-series using the fact that (a) $a_1^q, \ldots, a_d^q$ is an $\mathbb{F}_p$-basis for G_k/G_{k+1} where $q = p^{k-1}$ and (b) G_k/G_{k+1} is central in G/G_{k+1}.

Having defined a coordinate system for a uniform group, we can now get hold of a Lie algebra. There is a very appealing construction of the Lie algebra in which we define an addition and Lie bracket directly onto the group.

We can already see an increasingly abelian character in a pro-p group as we descend the lower p-series since G_n/G_{2n} is abelian. To get the additive structure on G we "pull up" this abelian structure on the layer G_n/G_{2n} as follows:

DEFINITION 10. Let G be a uniform group and $x, y \in G$. Define for $n \in \mathbb{N}$

$$x +_n y = (x^{p^n} y^{p^n})^{p^{-n}}.$$

Note that this pulling up is only possible in a well-defined way in a uniform group since then every element of G_{n+1} has a unique p^n-th root in G. In fact the operation $+_n$ just defines the group structure of G_{n+1} on G since the map $x \mapsto x^{p^{-n}}$ defines an isomorphism from G_{n+1} onto $(G, +_n)$.

DEFINITION 11. Let G be a uniform group and $x, y \in G$. Define

$$x + y = \lim_{n \to \infty} x +_n y.$$

THEOREM 5. *Let $G = \overline{\langle a_1, \ldots, a_d \rangle}$ be a uniform pro-p group of dimension d. Then $(G, +, \mathbb{Z}_p)$ is a free $\mathbb{Z}_p$-module on the basis $\{a_1, \ldots, a_d\}$ where the $\mathbb{Z}_p$ action is defined by $\lambda g = g^\lambda$.*

As an immediate corollary we get a faithful linear representation of degree d for $G/Z(G)$ by looking at the action of G on $(G, +, \mathbb{Z}_p)$ where G acts by conjugation and $Z(G)$ denotes the centre of G. We shall improve this result shortly when we consider the completed group algebra of G.

3. Example: the $\mathbb{Z}_p$-module of $\mathrm{GL}_n^1(\mathbb{Z}_p)$. Consider what this addition is in the group $G = \mathrm{GL}_n^1(\mathbb{Z}_p) = 1 + p\mathrm{M}_n(\mathbb{Z}_p)$. Take $1 + X$ and $1 + Y \in G$ and define $\log(1 + X) = \sum_{i=1}^{\infty} (-1)^i X^i / i = X' \in p\mathrm{M}_n(\mathbb{Z}_p)$ and similarly $\log(1 + Y) = Y'$. The function log in this non-commutative setting satisfies the following identities: if $X, Y \in p^i\mathrm{M}_n(\mathbb{Z}_p)$ then

$$\log(1 + X)^\lambda = \lambda \log(1 + X)$$

$$\log(1 + X)(1 + Y) = X' + Y' + p^{2i} Z$$

for some $Z \in \mathrm{M}_n(\mathbb{Z}_p)$. Applying this to our definition of addition above we get

$$\lim_{i\to\infty} \log\big((1+X)^{p^i}(1+Y)^{p^i}\big)^{p^{-i}} = \lim_{i\to\infty} \big(p^{-i}(p^iX' + p^iY' + p^{2i+2}Z_i)\big) = X' + Y'.$$

So we see that the addition that we have defined on the uniform group G actually picks up the addition on $G - 1 = \mathrm{M}_n(\mathbb{Z}_p)$. Note that this coordinate structure is not the same as that derived in Theorem 4 but it is a non-trivial fact that it is analytically equivalent to it, i.e. there are analytic functions getting you from one to the other. Since we have defined the additive structure on $\mathrm{M}_n(\mathbb{Z}_p)$, is it possible to define the associated Lie structure, namely $(1+X, 1+Y) = X'Y' - Y'X'$?

This time we use the following identity for log: if $X, Y \in p^i\mathrm{M}_n(\mathbb{Z}_p)$ then

$$\log[1+X, 1+Y] = X'Y' - Y'X' + p^{3i}Z.$$

We look to make the error term in this expression as small as possible. The trick is to pull up structure from lower down the group:

$$\begin{aligned}\lim_{i\to\infty} \log\big[(1+X)^{p^i}, (1+Y)^{p^i}\big]^{p^{-2i}} &= \lim_{i\to\infty} \big(p^{-2i}(p^iX'p^iY' - p^iY'p^iX' + p^{3i+3}Z_i)\big)\\ &= X'Y' - Y'X'.\end{aligned}$$

We can generalize this to any uniform pro-p group:

Definition 12. For $x, y \in G$, a uniform group, and $n \in \mathbb{N}$, define

$$(x, y)_n = [x^{p^n}, y^{p^n}]^{p^{-2n}}.$$

Note that $[x^{p^n}, y^{p^n}] \in [G_{n+1}, G_{n+1}] \leq G_{2n+2}$ so that we can take the p^{2n}-th root of this element. (In fact of course there exists a p^{2n+1}-th root but this does not define anything significantly different.)

Definition 13. For $x, y \in G$, a uniform group, define

$$(x, y) = \lim_{n\to\infty} (x, y)_n.$$

Proposition 10. *$\mathcal{L}(G) = \big(G, +, (\ ,\ )\big)$ is a $\mathbb{Z}_p$-Lie algebra.*

This proposition can be proved directly by checking axioms; but the easiest way is to identify $\mathcal{L}(G)$ inside an associative algebra which we construct below where addition and the Lie bracket defined on G correspond to addition and $XY - YX$ inside the algebra. This is what the map log did in the example $GL_n'(\mathbb{Z}_p)$.

What hope is there to generalise this analysis of a coordinate structure and construction of a Lie algebra in the case of a pro-p group analytic over $\mathbb{F}_p[\![T]\!]$? The key to the uniform pro-p case is the crucial role that the p-operator plays. Some recent work [**7**] due to David Cartwright begins an investigation into pro-p groups with an operator T which aims to imitate the action of multiplication by T in $\mathbb{F}_p[\![T]\!]$.

In his thesis [**7**], Cartwright shows that under certain axioms on an operator T on a pro-p group we can define an addition and Lie bracket as in the Definitions 11 and 13 above where we replace the action of p by that of T. If we consider the example $G = \mathrm{GL}_n^1\big(\mathbb{F}_p[\![T]\!]\big) = 1 + T\,\mathrm{M}_n\big(\mathbb{F}_p[\![T]\!]\big)$ where T sends $1+X$ to $1+TX$ then a similar but simpler argument to the example above will ensure that we still pick out the addition and Lie bracket $XY - YX$ on $T\,\mathrm{M}_n\big(\mathbb{F}_p[\![T]\!]\big)$. In fact this is a general phenomenon for a class of groups called standard groups which we shall define in the next section. Cartwright does not have recourse to a logarithm map and so is forced to prove directly that the identities for a Lie algebra are satisfied by the appropriate operations.

A crucial tool in proceeding any further with constructing the Lie theory of a uniform group is the associated completed group algebra. This is defined to be the inverse image of the group algebras over $\mathbb{Z}_p$ of finite images of G and is in fact defined for any pro-p group:

$$\Lambda(G) = \varprojlim_{N \text{ open in } G} \mathbb{Z}_p[G/N]$$

In the case where G is uniform we can tensor Λ with $\mathbb{Q}_p$ and inside this algebra we can define a logarithm and exponential map. The image in $\Lambda \otimes \mathbb{Q}_p$ of G under the logarithm map is in fact isomorphic to the Lie algebra $\mathcal{L}(G)$ of Proposition 10 where the Lie bracket corresponds to the Lie bracket $XY - YX$ in the associative algebra $\Lambda \otimes \mathbb{Q}_p$.

It is inside the completed group algebra that we can construct a faithful representation for G. Ado's Theorem guarantees us a faithful linear representation for the Lie algebra $\mathcal{L}(G) = \log(G)$. Using the exponential map, it is possible to pull this representation from $\mathcal{L}(G)$ over to a faithful representation of G.

THEOREM 6. *Let G be a uniform pro-p group. Then G admits a faithful linear representation $\phi \colon G \to \mathrm{GL}_n(\mathbb{Q}_p)$.*

The completed group algebra Λ plays a crucial role in Iwasawa theory where it is called the Iwasawa algebra. Classical Iwasawa theory concerns the case of cyclotomic field extensions with Galois group G isomorphic to $\mathbb{Z}_p^*$. In this case the Iwasawa algebra or completed group algebra Λ of G is isomorphic to the ring $\mathbb{Z}_p[\![T]\!]$. The reason this algebra is of interest to number theorists is that it can be identified with the set of p-adic valued measures on G. It is these measures which allow one to define p-adic L-functions in an algebraic way and in particular to identify naturally the Kubota-Leopoldt L-function. The key to this identification is the understanding we have of Λ-modules when $\Lambda = \mathbb{Z}_p[\![T]\!]$.

It would be of interest to generalize this approach for other non-commutative Galois groups, in particular to develop Iwasawa theory for $G = \mathrm{SL}_2(\mathbb{Z}_p)$, a Galois group which occurs naturally in elliptic curves. However the completed group algebra Λ of $\mathrm{SL}_2(\mathbb{Z}_p)$ and the structure of the corresponding Λ-modules is not at present very well understood. We do at least know that if G is a uniform pro-p group then its completed group algebra Λ is a Noetherian ring with no zero-divisors.

4. Groups Analytic over Local Rings

There is an accepted theory of groups analytic over a discrete valuation ring. However, we would like to widen this definition to encompass groups like

$$\mathrm{GL}_n(\mathbb{Z}_p[\![T_1, \ldots, T_m]\!])$$

which deserve to be called analytic over $\mathbb{Z}_p[\![T_1, \ldots, T_m]\!]$. Lubotzky and Shalev have begun to investigate groups that fit into this category called standard groups. We shall define standard groups in what follows. They look very like the congruence subgroups that we defined in $\mathrm{GL}_n(\mathbb{Z}_p[\![T_1, \ldots, T_m]\!])$. The theory of analytic groups over discrete valuation rings can be reduced to an analysis of standard groups since they sit as open subgroups inside such analytic groups. We are therefore looking for a definition of the category of groups analytic over a ring like $\mathbb{Z}_p[\![T_1, \ldots, T_m]\!]$ which will have the same property.

We extend the definition of analytic groups from discrete valuation rings to commutative local rings $(R, \mathfrak{m})$ complete with respect to the $\mathfrak{m}$-adic topology which are integral domains with K the corresponding field of fractions. For example $(\mathbb{Z}_p[\![T_1, \ldots, T_m]\!], \langle p, T_1, \ldots, T_m\rangle)$ is such a ring. The field of fractions is

$$\mathbb{Q}_p((T_1, \ldots, T_m)),$$

which consists of the set of finite tailed Laurent series $F = \sum_{\alpha \in \mathbb{Z}^m} c_\alpha X^\alpha$ where $c_\alpha = 0$ if (for some i) $\alpha_i < -N(F)$ for some $N(F) \in \mathbb{N}$. The notation we shall use is that for $\alpha \in \mathbb{N}^m$, $X^\alpha = X_1^{\alpha_1}, \ldots, X_n^{\alpha_n}$.

With this set up, it makes sense to evaluate a formal power series

$$F \in R[\![X_1, \ldots, X_n]\!]$$

at points of the maximal ideal $(x_1, \ldots, x_n) \in \mathfrak{m}^{(n)}$, since the maximal ideal consists of topologically nilpotent elements. We call the value $F(x_1, \ldots, x_n)$. [Note that we shall write $\mathfrak{m}^{(n)}$ for n-tuples with entries from $\mathfrak{m}$ to distinguish it from the power of the maximal ideal $\mathfrak{m}^n$.]

However we shall also want to consider evaluating

$$F = \sum_{\alpha \in \mathbb{N}^m} c_\alpha X^\alpha \in K[\![X_1, \ldots, X_n]\!].$$

Evaluating F at $(x_1, \ldots, x_n) \in \mathfrak{m}^{(n)}$ is well defined provided that for all N there exists M such that $c_\alpha x^\alpha \in \mathfrak{m}^N$ for all $\langle \alpha \rangle = \alpha_1 + \cdots + \alpha_n \geq M$. Whenever we write $F(x)$ we shall assume that this convergence criteria is satisfied and that the sum is $F(x)$.

With this definition of evaluating $F(X) \in K[\![X_1, \ldots, X_n]\!]$ we can build up our definition of a group analytic over R. We begin with the definition of an R-analytic manifold.

DEFINITION 14. (1) Let $U \subset \mathfrak{m}^{(n)}$ be open and let $\phi \colon U \to K$ be a function. Then ϕ is said to be *analytic* in U if for each $x \in U$ there is a formal power series $F \in K[\![X_1, \ldots, X_n]\!]$ and $N > 0$ such that
   (i) $x + (\mathfrak{m}^N)^{(n)} \subset U$ and
   (ii) F converges in $(\mathfrak{m}^N)^{(n)}$ and for $y \in (\mathfrak{m}^N)^{(n)}$, $\phi(x + y) = F(y)$.

(2) Let $U \subset \mathfrak{m}^{(n)}$ be open and let $\phi = (\phi_1, \ldots, \phi_m) \colon U \to K^{(m)}$ be a function. Then ϕ is said to be *analytic* in U if each ϕ_i is analytic for $i = 1, \ldots, m$.

(3) Let X be a topological space. An *R-chart* c on X is a triple $c = (U, \phi, n)$ such that
   (i) U is open in X,
   (ii) $n \in \mathbb{N}$ and
   (iii) $\phi : U \to \phi U \subset \mathfrak{m}^{(n)}$ is open and ϕ is a homeomorphism.

(4) Two R-charts $c = (U, \phi, n)$ and $c' = (U', \phi', n')$ are said to be *compatible* if setting $V = U \cap U'$, the maps $\phi' \circ \phi^{-1}|_{\phi(V)}$ and $\phi \circ \phi'^{-1}|_{\phi'(V)}$ are analytic.

(5) An *R-atlas* on a topological space X is a set

$$A = \{(U_i, \phi_i, n_i) : i \in I\}$$

with the properties that
   (i) for each $i \in I$, (U_i, ϕ_i, n_i) is an R-chart on X,
   (ii) for all $i, j \in I$, (U_i, ϕ_i, n_i) and (U_j, ϕ_j, n_j) are compatible, and
   (iii) $X = \bigcup_{i \in I} U_i$.

Two atlases are compatible if all their charts are compatible.

(6) An *analytic manifold structure over* R on a topological space X is an equivalence class of compatible atlases.

1. Examples: analytic manifolds over *R*.

EXAMPLE 19. Let $X = K^{(m)}$ where K is the field of fractions of the local ring $(R, \mathfrak{m})$. Let $\mathfrak{m} = \langle T_1, \ldots, T_n \rangle$. Put $U_i = T_1^{-i_1} \ldots T_n^{-i_n} \mathfrak{m}^{(m)}$ for each $i = (i_1, \ldots, i_n) \in \mathbb{N}^n$. Then we can write $K = \bigcup_{i \in \mathbb{N}^n} U_i$. We can make X into an analytic manifold structure over R by defining the atlas $A = \{(U_i, \phi_i, n_i) : i \in \mathbb{N}^n\}$ where $\phi_i \colon U_i \to \mathfrak{m}^{(m)}$ is defined by $\phi_i(u) = T_1^{i_1} \ldots T_n^{i_n} u$. These charts are compatible since $\phi_i \circ \phi_j^{-1}|_{\phi_j(U_i \cap U_j)}(u) = T_1^{i_1 - j_1} \ldots T_n^{i_n - j_n} u$.

Note that had we tried to define compatibility just using power series with coefficients in R these charts would not be compatible.

EXAMPLE 20. Let $X = \mathrm{GL}_n(K)$ then X has a manifold structure over R defined as follows: Let $U = 1_n + \mathrm{M}_n(\mathfrak{m})$, the congruence subgroup inside $\mathrm{GL}_n(R)$ which is an open subgroup of X. Define $\phi \colon U \to \mathrm{M}_n(\mathfrak{m})$ by $\phi(u) = (u - 1_n)$ for all $u \in U$. For each $h \in X$, let $V_h = hU$, an open neighbourhood of h in X. Define $\phi_h \colon V_h \to \mathrm{M}_n(\mathfrak{m})$ by $\phi_h(x) = \phi(h^{-1}x)$. Then $A = \{(V_h, \phi_h, n^2) : h \in X\}$ defines an R-atlas on X.

EXAMPLE 21 (Uniform group). We have already defined a $\mathbb{Z}_p$-manifold structure on a uniform group $G = \overline{\langle a_1, \ldots, a_d \rangle}$ in Theorem 4. If $x \in G$ there exist unique $\lambda_1, \ldots, \lambda_d \in \mathbb{Z}_p$ such that

$$x = a_1^{\lambda_1} \cdots a_d^{\lambda_d}.$$

The map $\phi \colon G \to \mathbb{Z}_p^d$ defined by $x \mapsto (\lambda_1, \ldots, \lambda_d)$ gives a chart and hence a manifold structure on the whole of G. There is an alternative chart that we can define using the intrinsic Lie algebra structure. Since $(G, +, \mathbb{Z}_p)$ is a free $\mathbb{Z}_p$-module, we can identify G with $\mathbb{Z}_p^d$ giving us another chart. It is a non-trivial fact that these two charts are in fact compatible and hence define the same manifold structure. At the heart of this fact is an application of the Baker-Campbell-Hausdorff formula which relates the group structure and Lie algebra structure of a Lie group.

Having defined the concept of an R-analytic manifold, we can now define an R-analytic group. To do this we need the concept of an analytic function between two manifolds:

DEFINITION 15. Let X and Y be two R-analytic manifolds. A function $f \colon X \to Y$ is called an *analytic function* if there exist R-atlases A and B of X and Y respectively such that, for all $(U, \phi, m) \in A$ and $(V, \psi, n) \in B$, setting $W = U \cap f^{-1}V$, the function $\psi \circ f \circ \phi^{-1} \colon \phi W \to \mathfrak{m}^{(n)}$ is an analytic function on the open subset ϕW of $\mathfrak{m}^{(n)}$.

DEFINITION 16. Let G be a topological group with the structure of an R-analytic manifold. We call G an R-analytic group if the group operations are analytic functions with respect to this manifold structure.

2. Examples: R-analytic groups.

EXAMPLE 22. Let $G = (K^{(m)}, +)$ where K is the field of fractions of the local ring $(R, \mathfrak{m})$. Then G is an R-analytic group with respect to the R-manifold defined in Example 19.

EXAMPLE 23. Let $G = (K^*, \cdot)$ the multiplicative group of the field. The Example 20 with $n = 1$ defines a manifold structure on this group. If $1 + u \in 1 + \mathfrak{m}$ then $\phi\big((1+u)^{-1}\big) = \sum_{n=1}^{\infty}(-1)^n u^n$ which is analytic in $\phi(1+u)$. In general, if $h \in G$ and $u \in \mathfrak{m}$ then $\phi_{h^{-1}}\big(\big(h(1+u)\big)^{-1}\big) = \sum_{n=1}^{\infty}(-1)^n u^n$ which is analytic in $\phi_h\big(h(1+u)\big) = u$. Multiplication is clearly analytic with respect to this manifold. Hence G is an R-analytic group.

EXAMPLE 24. We can generalize the argument in Example 23 to show that $G = \mathrm{GL}_n(K)$ with the R-analytic structure defined in Example 20 is an R-analytic group.

EXAMPLE 25 (Uniform group). Lazard showed using the completed group algebra Λ of a uniform group that the group operations are analytic with respect to Example 21 of an R-analytic manifold. Is there any possibility for a more direct proof by successively approximating the group operation as we descend the lower p-series? On the first layer, the group operation is given by addition of coordinates. Such a direct approach might be very helpful in trying to prove analogues of Lazard's result for other rings R.

We define now an important class of R-analytic groups called R-standard groups. The properties of these groups have been considered by the Bourbaki school [**5**] for discrete valuation rings and more generally by Lubotzky and Shalev in [**26**]. In the case of discrete valuation rings R, analytic groups over R were shown to contain an open standard group. So in some sense analytic groups are close to being standard groups.

Lubotzky and Shalev saw that many of the arguments about the properties of standard groups work perfectly well even if R is not a discrete valuation ring. These standard groups over R certainly deserved to be examples of R-analytic groups however no such category had been defined in their paper. We hope that this present paper along with the account in [**10**] will help to put Lubotzky and Shalev's work into a suitable framework. [I should make some comment on the decision to use the name R for the pro-p ring rather then Lubotzky and Shalev's choice of Λ. This is to avoid confusion with the completed group algebra or Iwasawa algebra which is traditionally denoted by number theorists by Λ.]

The key test of our definition of R-analytic groups will be whether we can prove that they always contain open R-standard subgroups. However to do this we shall need to broaden the definition of a standard group.

We begin with the following:

DEFINITION 17. A *formal group law* in d variables over a ring R is a d-tuple $f(X,Y) = \big(f_i(X,Y)\big)$ where $f_i \in R[\![X_1,\dots,X_d,Y_1,\dots,Y_d]\!] = R[\![X,Y]\!]$ with the property that

(1) $f(X,0) = f(0,X) = X$; and
(2) $f\big(X, f(Y,Z)\big) = f\big(f(X,Y),Z\big)$.

Notice that (1) implies that the constant terms of the power series f_i are all 0. So (2) makes sense since substitution of power series $f\big(g(X)\big)$ can only be affected if g has zero constant term.

3. Examples: formal group laws.

EXAMPLE 26. $f(X,Y) = X + Y$ is a formal group law defining the additive structure of the ring.

EXAMPLE 27. $f(X,Y) = X+Y+XY$ is a formal group law which defines the multiplicative structure on R if one observes that $X+Y+XY = (1+X)(1+Y)-1$.

EXAMPLE 28. We can generalize this to define a formal group law which tells us how to define matrix multiplication in $1+M_n(\mathfrak{m})$. Let $X = (X_{ij})$ and $Y = (Y_{ij})$ define $n \times n$ matrices then $f(X,Y) = X+Y+X \cdot Y$ defines a formal group law.

We have seen that the power series in a formal group law converge on the maximal ideal $\mathfrak{m}$ and take values in $\mathfrak{m}$. Hence we can try to use a formal group law of dimension n to define a group operation on the set $\mathfrak{m}^{(n)}$. The definition of a formal group has been set up to ensure that this operation really does define a group operation. But something still needs to checked: namely that identities in formal power series translate into the corresponding identity for our group operation when evaluated on $\mathfrak{m}$.

This is a non-trivial point as it is possible that both sides of a formal power series can converge at some point but have different values. For example, in $\mathbb{Q}_2$ the formal power series corresponding to $\log(X)$ converges at $X = -1$ to 0 and $\exp(0) = 1$. Hence the formal power series identity $\exp \circ \log(X) = X$ does not translate to an identity when evaluated at $X = -1$.

However provided that we stay within the ideal $\mathfrak{m}$ we don't get this sort of pathology. This was proved in Theorem 7.25 of [**8**] for $\mathbb{Q}_p$-algebras. For our more general setting we can refer to Corollary of Proposition 6 of III §4.5 in [**4**] which states the following:

PROPOSITION 11. *Let A be a commutative, linearly topologized, Hausdorff, complete ring containing a closed ideal $\mathfrak{m}$ of topologically nilpotent elements. Let $g = (g_1, \ldots, g_q)$ be a system without constant term of series in $A[\![X_1, \ldots, X_n]\!]$ and $f = (f_1, \ldots, f_r)$ a system of series in $A[\![X_1, \ldots, X_q]\!]$. If $x = (x_1, \ldots, x_n) \in \mathfrak{m}^{(n)}$ then*

$$(f \circ g)(x) = f\big(g(x)\big)$$

i.e. evaluation and substitution of power series commute on $\mathfrak{m}$.

DEFINITION 18. Let $(R, \mathfrak{m})$ be our local ring and f be a formal group law of dimension d. Then $G = \big(\mathfrak{m}^{(d)}, f\big)$ is a group where $xy = f(x,y) \in \mathfrak{m}^{(d)}$ defines the group operation on $\mathfrak{m}^{(d)}$. Such a group is called an R-standard group. Define for each $n \in \mathbb{N}$

$$G_n = \big((\mathfrak{m}^n)^{(d)}, f\big).$$

This is a group which we call an *R-standard group of level n*.

If $G = \big(\mathfrak{m}^{(d)}, f\big)$ is a standard group then the chart $\big(\mathfrak{m}^{(d)}, \phi, d\big)$ where ϕ is the identity map, defines an R-analytic manifold structure on G which makes G into an R-analytic group.

4. Examples: *R*-standard groups.

EXAMPLE 29. $G = \big(\mathfrak{m}^{(d)}, +\big)$ is an R-standard group.

EXAMPLE 30. $G = (1+\mathfrak{m}, \cdot)$ is an R-standard group defined by the formal group law $X+Y+XY$.

EXAMPLE 31. Let $\mathrm{GL}_n^1(R) = 1 + \mathrm{M}_n(\mathfrak{m})$ be the congruence subgroup inside $\mathrm{GL}_n(R)$, then $\mathrm{GL}_n^1(R)$ is an R-standard group defined by Example 28 of a formal group law $(X_{ij}) + (Y_{ij}) + (X_{ij}) \cdot (Y_{ij})$.

EXAMPLE 32. Let G be a uniform pro-p group of dimension d. Then we have identified G with $\mathbb{Z}_p^{(d)}$ where the group operation is defined by power series with coefficients in $\mathbb{Q}_p$. So a uniform group is not in general a $\mathbb{Z}_p$-standard group. However there exists some term of the lower p-series $G_n = p^{n-1}\mathbb{Z}_p^{(d)}$ such that when we identify it with $p\mathbb{Z}_p^{(d)}$ the group operation is defined by power series with coefficients in $\mathbb{Z}_p$, making G_n into a $\mathbb{Z}_p$-standard group.

The next theorem shows that as for analytic groups over discrete valuation rings, the theory of R-analytic groups as defined above reduces to the study of R-standard groups:

THEOREM 7. *Let $(R, \mathfrak{m})$ be a complete, local, commutative Noetherian, integral domain with field of fractions K and finite residue field $R/\mathfrak{m}$. Let G be an R-analytic group. Then G contains an open R-standard group of level n for some $n \in \mathbb{N}$.*

Proof. There is a dichotomy in the proof between rings of Krull dimension greater than 1 and the classical case of rings of Krull dimension 1, i.e. discrete valuation rings.

Let A be an atlas defining the manifold structure on G. There exists a chart $(U, \phi, n) \in A$ with $1 \in U$. Since G is an analytic group, the function $f\colon G \times G \to G$ defined by $(x, y) \mapsto xy^{-1}$ is analytic. Hence there exists some neighbourhood $\phi(1) + (\mathfrak{m}^h)^{(n)}$ of the point $\phi(1)$ and power series $F_j \in K[\![X, Y]\!]$ $(j = 1, \ldots, n)$ such that for all λ and $\mu \in (\mathfrak{m}^h)^{(n)}$

$$\phi_j \circ f \circ (\phi \times \phi)^{-1}(\phi(1) + \lambda, \phi(1) + \mu) = F_j(\lambda, \mu).$$

By a simple translation of coordinates we can assume that $\phi(1) = 0$ in $K^{(n)}$ since this defines a compatible chart. The power series F_j then look like $F_j(X, Y) = X_j - Y_j + \sum_{\alpha,\beta\in\mathbb{N}^n} c_{j,\alpha,\beta} X^\alpha Y^\beta$. (To deduce this form requires applying the fact that two power series which take the same values on some open subset of $K^{(n)}$ are identical.) If the coefficients $c_{j,\alpha,\beta} \in R$ for all $\alpha, \beta \in \mathbb{N}^n$ then we would be home since $\big(\phi^{-1}\big((\mathfrak{m}^h)^{(n)}\big), F\big)$ would be an R-standard subgroup of level h. Note that it is a subgroup since F evaluated on $(\mathfrak{m}^h)^{(n)}$ will take values in $(\mathfrak{m}^h)^{(n)}$ if $c_{j,\alpha,\beta} \in R$. However in general the coefficients need not be integral.

Suppose that R is a discrete valuation ring with maximal ideal $\mathfrak{m}$ generated by the single element π. In this case we can rescale our chart using powers of π to get the coefficients to be integral. With respect to the new chart $x \mapsto \pi^{-k}\phi(x)$ the group operation is given by power series $F_j'(X, Y) = X_j - Y_j + \sum_{\alpha,\beta\in\mathbb{N}^n} c_{j,\alpha,\beta}\pi^{(\langle\alpha\rangle+\langle\beta\rangle-1)k} X^\alpha Y^\beta$. Why can we choose k so that $c_{j,\alpha,\beta}\pi^{(\langle\alpha\rangle+\langle\beta\rangle-1)k} \in R$ for all $\alpha, \beta \in \mathbb{N}^n \setminus 0$? This follows from the fact that, since F_j converge on $(\mathfrak{m}^h)^{(n)}$, $c_{j,\alpha,\beta}\pi^{(\langle\alpha\rangle+\langle\beta\rangle)h} \in R$ for $\langle\alpha\rangle + \langle\beta\rangle \geq N$ for some N large enough. Hence for some l we can get $c_{j,\alpha,\beta}\pi^{(\langle\alpha\rangle+\langle\beta\rangle)(l+h)} \in R$ for all $\alpha, \beta \in \mathbb{N}^n \setminus 0$. Now choose $k = 2(l + h)$. Hence the open subset $H = \phi^{-1}\big((\pi^{k-h+1}\mathfrak{m}^h)^{(n)}\big) = \phi^{-1}\big((\mathfrak{m}^{(k+1)})^{(n)}\big)$ is in fact an R-standard group of level 1 since with respect to the chart $x \mapsto \pi^{-k}\phi(x)$ the group H looks like $(\mathfrak{m})^{(n)}$ with the group operation given by power series $F_j'(X, Y) = X_j - Y_j + \sum_{\alpha,\beta\in\mathbb{N}^n} c_{j,\alpha,\beta}\pi^{(\langle\alpha\rangle+\langle\beta\rangle-1)(k+h)} X^\alpha Y^\beta$.

If the maximal ideal $\mathfrak{m}$ is in fact generated by two elements, e.g. $R = \mathbb{Z}_p[\![T]\!]$ and $\mathfrak{m} = \langle p, T\rangle$ then at first sight it appears that we can do the same analysis to rescale the power series using powers of p and T. However when we come to consider the subset $H = \phi^{-1}\big((T^a p^b \mathfrak{m}^h)^{(n)}\big)$ this is no longer an open subset of G.

However, although one hand take away, the other hand give . In a discrete valuation ring, a power series can converge on $\mathfrak{m}$ but have all its coefficients not integral, e.g. the exponential function. It turns out that this cannot happen in rings like $R = \mathbb{Z}_p[\![T]\!]$: if a power series converges on some power of the maximal ideal $\mathfrak{m} = \langle p, T\rangle$ taking values in R, then this is a much stronger condition forcing the coefficients to be in R. We prove this for the particular example of $R = \mathbb{Z}_p[\![T]\!]$ and refer the reader to [**10**] for a proof of the general case.

LEMMA 3. *Let* $F(X) = \sum_{\alpha\in\mathbb{N}^n} a_\alpha X^\alpha \in K[\![X]\!]$ *where* $K = \mathbb{Q}_p((T))$, *the field of fractions of* $R = \mathbb{Z}_p[\![T]\!]$. *Suppose that* F *converges on some power* $\mathfrak{m}^h$ *of the maximal ideal* $\mathfrak{m} = \langle p, T\rangle$ *and takes values in* $R = \mathbb{Z}_p[\![T]\!]$, *i.e.* $F(x) \in R$ *for all* $x \in (\mathfrak{m}^h)^{(n)}$. *Then* $a_\alpha \in R$ *for all* $\alpha \in \mathbb{N}^n$.

PROOF. We exploit the fact that there is more room for manoeuvre in R with two degrees of freedom. We first prove that this is true for almost all $\alpha \in \mathbb{N}^n$. Suppose not, then there are two cases to consider.

CASE 1. For infinitely many $\alpha \in I \subseteq \mathbb{N}^n$ say, $a_\alpha = T^{-n(\alpha)}b_\alpha$ with $n(\alpha) > 0$ where $b_\alpha = c_0 + c_1T + c_2T^2 + \cdots$ with $c_0 \neq 0$. But F converges at p^h, hence for almost all α, $a_\alpha p^{\langle\alpha\rangle h} \in \mathfrak{m}$. But for $\alpha \in I$, $a_\alpha p^{\langle\alpha\rangle h} = T^{-n(\alpha)}b_\alpha p^{\langle\alpha\rangle h} \notin \mathfrak{m}$.

CASE 2. For infinitely many $\alpha \in I \subseteq \mathbb{N}^n$ say, $a_\alpha = p^{-n(\alpha)}b_\alpha$ with $n(\alpha) > 0$ where $b_\alpha = \lambda T^m + c_\alpha$ with $\lambda \in \mathbb{Z}_p^*, m \in \mathbb{Z}$ and $c_\alpha \in \mathbb{Z}_p((T))$. But F converges at T^h, hence for almost all α, $a_\alpha T^{\langle\alpha\rangle h} \in \mathfrak{m}$. But for $\alpha \in I$, $a_\alpha T^{\langle\alpha\rangle h} = p^{-n(\alpha)}b_\alpha T^{\langle\alpha\rangle h} \notin \mathfrak{m}$.

We can now reduce to the case where $F(X)$ is a polynomial by subtracting the power series consisting of all the terms which already have integral coefficients. In fact we can assume also then that the polynomial is in a single variable X. So we can write

$$F(X) = p^{-r}T^{-s}\sum_{i=0}^{m} a_i X^i$$

where $a_i = \sum_{j\in\mathbb{N}} a_{i,j}T^j \in \mathbb{Z}_p[\![T]\!]$. We may assume that r and s have been chosen minimally so that $a_{i_0,0} \neq 0$ for some i_0 and $a_{i,j} \notin p\mathbb{Z}_p$ for some i, j. Our task is to prove that $r = s = 0$ by canny choices of $X = T^e p^f$ where $e + f \geq h$. The coefficient of T^q in $F(T^e p^f)$ is

$$\sum_{i=0}^{m} p^{if-r} a_{i,q+s-ie} \begin{cases} \in \mathbb{Z}_p \\ = 0 \text{ for } q < 0 \end{cases}$$

since F takes values in $R = \mathbb{Z}_p[\![T]\!]$. We show first that if $s > 0$ then $a_{i,0} = 0$ for all $i = 0, \ldots, m$. For each $l \geq 1$ take $f = l \cdot h$, $e = 0$, and look at the coefficient of T^q for $q = -s$. Then since $q < 0$ we get infinitely many expressions for each choice of l

$$a_{0,0} + p^{lh}a_{1,0} + p^{2lh}a_{2,0} + \cdots + p^{mlh}a_{m,0} = 0.$$

Hence $a_{i,0} = 0$ for all $i = 0, \ldots, m$. But, by our minimal choice of s, $a_{i_0,0} \neq 0$ for some i_0. Hence $s = 0$.

We now attack the powers of p. We show that $a_{i,j} \in p^r\mathbb{Z}_p$ for all i, j. Note that $a_{i,j} = 0$ if $j < 0$. Choose $e = (m+1)\cdot h, f = 0$. For $q = j$ we get

$$p^{-r}a_{0,j} \in \mathbb{Z}_p. \tag{1}$$

Consider next $q = e + i$ which gives $p^{-r}a_{0,j} + p^{-r}a_{1,j} \in \mathbb{Z}_p$. By (1) we get that

(2) $$p^{-r}a_{1,j} \in \mathbb{Z}_p.$$

We can continue inductively in this fashion by choosing $q = ie + j$ to deduce that $p^{-r}a_{i,j} \in \mathbb{Z}_p$ for all i, j. By our minimal choice of r $a_{i,j} \notin p\mathbb{Z}_p$ for some i, j. Hence $r = 0$.

This proves our lemma. □

Thus the situation for rings of Krull dimension greater than 1 is actually much simpler than for a discrete valuation ring since the power series F_j will already have integral coefficients and so we are already home with the argument given in the first paragraph of the proof.

Note that in a discrete valuation ring R it is possible to scale an R-standard group of level h to look like an R-standard group of level 1, i.e. the more conventional definition of a standard group. However for rings R of Krull dimension greater than 1, this is not possible and we need this new distinction between standard groups of different levels.

These R-standard subgroups of level n, $G_n = \left((\mathfrak{m}^n)^{(d)}, f\right)$ for $n \in \mathbb{N}$, or "congruence" subgroups of an R-standard group define the topology on an R-standard group G revealing G to be a pro-p group:

LEMMA 4. *Let G be an R-standard group*

(1) *G_n are normal subgroups and form a base of neighbourhoods for the identity in G.*
(2) *$G_n/G_{n+1} \cong \left((\mathfrak{m}^n/\mathfrak{m}^{n+1})^{(d)}, +\right)$, a finite p-group.*
(3) *G is a pro-p group.*

How close is this filtration to being the lower p-series of G? In Examples 16 and 17 we saw that if G is the R-standard group $\mathrm{GL}_n^1(\mathbb{Z}_p)$ or $\mathrm{SL}_n^1(\mathbb{F}_p[\![T]\!])$ then the congruence kernels G_n coincide with the lower p-series. This is not in general the case. The following Lemma describes what can be said in general:

LEMMA 5. *Let G be an R-standard group. For all n and $m \in \mathbb{N}$*

(1) *$[G_m, G_n] \leq G_{m+n}$.*
(2) *$G_n \geq \gamma_n(G)$ where $\gamma_n(G)$ is the n-th term of the lower central series of G.*
(3) *If $R = \mathbb{Z}_p$ then $G_n^p = G_{n+1}$ and $G_n = P_n(G)$, the lower p-series of G.*
(4) *If $pR = 0$ (e.g. $R = \mathbb{F}_p[\![T]\!]$) then $G_n^p \leq G_{pn}$.*

PROOF. Let f be the formal group law defining the R-standard group G. Let $O(n)$ denote power series in which the homogeneous terms all have degree at least n and $O'(n)$ is those power series in $O(n)$ in which some X_i and some Y_j occur in each homogeneous component. For example X^2, Y^2 and $XY \in O(2)$ but only $XY \in O'(2)$. The defining properties of a formal group law imply the following (see **[8]** Chapter 9, Section 4):

$$f = X + Y + B(X, Y) + O'(3)$$
$$[X, Y] = B(X, Y) - B(Y, X) + O'(3)$$
$$f^p = pX + O(2) \in O(p) \mod p$$

where $B(X, Y)$ is a bilinear form in X and Y, $[X, Y]$ denotes the formal power series which corresponds to the commutator and f^p is the formal power series got by substituting f in itself successively p times, representing taking p-th powers. For

TABLE 1

	$\mathbb{Z}_p$-standard groups	$\mathbb{F}_p[\![T]\!]$-standard groups
(1)	Always finitely generated.	$\mathbb{F}_p[\![T]\!]$, $\mathbb{F}_p[\![T]\!]^*$, $\mathrm{GL}_n^1(\mathbb{F}_p[\![T]\!])$ are not finitely generated, but $\mathrm{SL}_n^1(\mathbb{F}_p[\![T]\!])$ is finitely generated.
(2)	G_n is the lower p-series.	$G_n \geq \gamma_n$ in general. For $\mathrm{SL}_n^1(\mathbb{F}_p[\![T]\!])$, $G_n = \gamma_n = P_n(G)$.
(3)	If G is both R-standard and $\mathbb{Z}_p$-standard then R is a finitely generated $\mathbb{Z}_p$-module.	Not clear.
(4)	dimension $= d(G)$.	Not in general, e.g. $(T\mathbb{F}_p[\![T]\!], +) \cong (T\mathbb{F}_p[\![T]\!]^{(2)}, +)$. However, $\dim(\mathrm{SL}_n^1(\mathbb{F}_p[\![T]\!]))$ $= n^2 - 1 = d(G)$.
(5)	Linear over $\mathbb{Q}_p$.	Unknown.
(6)	If G has finite rank or G is virtually powerful or $a_n(G)$ grows polynomially then G contains a $\mathbb{Z}_p$-standard group.	Unknown.
(7)	Closed subgroups are virtually $\mathbb{Z}_p$-standard.	Not clear, e.g. the closure of the infinite cyclic subgroup generated by $\begin{pmatrix} 1+T & 0 \\ 0 & (1+T)^{-1} \end{pmatrix}$ in $\mathrm{SL}_n^1(\mathbb{F}_p[\![T]\!])$ is isomorphic to $\mathbb{Z}_p$.

example, if $f = X+Y+XY$ then $f^p = (1+X)^p - 1$. By Proposition 11 the formal power series $[X, Y]$ and f^p when evaluated on G actually define commutation and taking p-th powers in the standard group G. Hence (1) follows from the structure of $[X, Y]$ and (2) is a corollary of (1). The structure of f^p yields (4) and, with a little more work (see [**8**] Theorem 9.33), (3). □

We raise a number of questions about these R-standard groups:

QUESTIONS. (1) When is an R-standard group finitely generated?
(2) What is the lower p-series or lower central series of an R-standard group?
(3) Can a pro-p group have standard structures over different rings R?
(4) Is the dimension of an R-standard group well-defined?
(5) Are R-standard groups linear over R?
(6) Can we tell if a finitely generated pro-p group G admits an R-standard structure on some open subgroup?
(7) What can we say about the structure of closed subgroups of R-standard groups?

The table above records what we know about these questions in the case of $\mathbb{Z}_p$-standard groups (a lot) and $\mathbb{F}_p[\![T]\!]$-standard groups (not much).

In (6) $a_n(G)$ denotes the following invariant

$$a_n(G) = \mathrm{card}\{H \leq G : |G{:}H| = n\}.$$

The group theoretic characterizations of p-adic analytic groups mentioned in (6) can be found in [**19**], [**24**] and [**25**].

One of the key strengths of the theory of Lie groups is the interplay between the group and its Lie algebra. In the case of Lie groups over $\mathbb{Z}_p$, this interplay amounts to an equivalence of categories between p-adic Lie groups and $\mathbb{Z}_p$-Lie algebras. We indicated that the logarithm and exponential maps inside the completed group algebra provide very useful tools for passing between the group and the Lie algebra. This is the key to proving for example the linearity of p-adic Lie groups.

We have seen how to construct the Lie algebra of a uniform group directly onto the group. There is an alternative way to realise this Lie algebra if the group has an R-standard structure which works for general rings R, and not just $\mathbb{Z}_p$.

Suppose that f is a formal group law defining some R-standard group G. As we have seen above f has the following form

$$f = X + Y + B(X, Y) + O'(3).$$

We can use the bilinear form $B(X, Y)$ to define a Lie bracket directly onto the standard group G.

DEFINITION 19. Set $(X, Y) = B(X, Y) - B(Y, X)$. Then $\mathcal{L}(G) = \big(\mathfrak{m}^{(d)}, +, (\ ,\ )\big)$ is an R-Lie algebra.

If $R = \mathbb{Z}_p$, then G is a uniform group and the Lie algebra we constructed in Proposition 10 directly on G is the same as this Lie algebra. We mentioned that Cartwright had imitated the direct construction of a Lie algebra in the case of a group with a T-operator. The canonical example of such a group is an $\mathbb{F}_p[\![T]\!]$-standard group where the T-operator is simply multiplication by T on the coordinates of the $\mathbb{F}_p[\![T]\!]$-standard group. The same proof as for $\mathbb{Z}_p$ shows that Cartwright's Lie algebra coincides in this case with the Lie algebra of Definition 19.

However unlike the p-adic case, the Lie algebra of an $\mathbb{F}_p[\![T]\!]$-standard group does not at present provide much information about the group. We no longer have the luxury of the logarithm and exponential functions in the characteristic p setting. So for example we can construct Lie algebras which do not arise as the Lie algebra of an $\mathbb{F}_p[\![T]\!]$-standard group. See for example [**30**] Exercise 1 of Chapter V, Part II.

As Table 1 demonstrates the general case of $\mathbb{F}_p[\![T]\!]$-standard groups is still little understood and in some cases looks rather unpromising for a sensible theory when there does not seem to be any dialogue between the group and its Lie algebra, nor is there a unique standard structure on a group and even the dimension is not well-defined. However it does seem that the case of $\mathrm{SL}_n^1\big(\mathbb{F}_p[\![T]\!]\big)$ offers rather more hope. Lubotzky and Shalev identified what it is that is making this example tick more sweetly, namely the fact pointed out in Example 17 that the Lie algebra $\mathfrak{sl}_n(\mathbb{F}_p)$ is perfect.

They exploited to good effect a Lie algebra which at first sight looks even weaker than the Lie algebra we have defined above. Using this Lie algebra they define a class of standard groups which behave in a similar fashion to $\mathrm{SL}_n^1\big(\mathbb{F}_p[\![T]\!]\big)$ called R-perfect standard groups.

DEFINITION 20. Let G be an R-standard group. Set

$$L_n = L_n(G) = G_n/G_{n+1} \cong \big((\mathfrak{m}^n/\mathfrak{m}^{n+1})^{(d)}, +\big)$$

for $n \in \mathbb{N}$. We define the *graded Lie algebra of G* to be

$$L(G) = \bigoplus_{n \geq 1} L_n(G)$$

where the Lie bracket is defined by

$$(xG_{n+1}, yG_{m+1}) = [x,y]G_{n+m+1} \\ = (x,y)_{\mathcal{L}(G)}G_{n+m+1}$$

where $(x,y)_{\mathcal{L}(G)}$ is the Lie bracket we defined on $\mathcal{L}(G)$. This second equality follows from the structure of the power series defining commutation.

This graded Lie algebra $L(G)$ is actually determined by a finite dimensional Lie algebra over the finite field $R/\mathfrak{m}$. Let $\overline{B}(X,Y) = B(X,Y) \mod \mathfrak{m}$ and define a new Lie bracket $\overline{(X,Y)} = \overline{B}(X,Y) - \overline{B}(Y,X)$ on the additive group $(R/\mathfrak{m})^{(d)}$. This defines a d-dimensional Lie algebra over the finite field $R/\mathfrak{m}$ which we call

$$L_0 = L_0(G) = \big((R/\mathfrak{m})^{(d)}, +, \overline{(X,Y)}\big).$$

Then we have the following description of the graded Lie algebra $L(G)$ in terms of this finite Lie algebra:

LEMMA 6. *$L(G) \cong L_0(G) \otimes_{R/\mathfrak{m}} \mathrm{gr}(\mathfrak{m})$ where $\mathrm{gr}(\mathfrak{m}) = \bigoplus_{n\geq 1} \mathfrak{m}^n/\mathfrak{m}^{n+1}$, the maximal ideal of the graded ring $\mathrm{gr}(R) = \bigoplus_{n\geq 0} \mathfrak{m}^n/\mathfrak{m}^{n+1}$.*

$L(G)$ therefore does not retain much information about the group G. For example, the graded Lie algebra of the R-standard group $\mathrm{SL}_n^1(R)$ has the following structure:

$$L\big(\mathrm{SL}_n^1(R)\big) \cong \mathfrak{sl}_n(R/\mathfrak{m}) \otimes_{R/\mathfrak{m}} \mathrm{gr}(\mathfrak{m}).$$

In particular, since $\mathrm{gr}(\mathbb{Z}_p) \cong \mathrm{gr}\big(\mathbb{F}_p[\![T]\!]\big)$ we have that

$$L\big(\mathrm{SL}_n^1(\mathbb{Z}_p)\big) \cong L\big(\mathrm{SL}_n^1(\mathbb{F}_p[\![T]\!])\big).$$

Referring back to Example 9, this is an example of a loop algebra and justifies calling $\mathrm{SL}_n^1\big(\mathbb{F}_p[\![T]\!]\big)$ the positive part of a loop group.

However Lubotzky and Shalev recognised that despite this loss of information, the graded Lie algebra could be put to good use. The observation that the structure of $\mathrm{SL}_n^1\big(\mathbb{F}_p[\![T]\!]\big)$ is well-behaved because of the perfectness of the Lie algebra $\mathfrak{sl}_n(\mathbb{F}_p)$ can now be generalised to the following class of groups:

DEFINITION 21. Let G be an R-standard group. We call G an *R-perfect group* if $L_0(G)$ is a perfect Lie algebra, i.e. $[L_0, L_0] = L_0$.

Some caution should be exercised with this definition since G itself is never perfect since it is a pro-p group with $[G,G] \leq G_2$.

Shalev and Lubotzky use exactly the same argument as for Example 17 of the calculation of the lower p-series of $\mathrm{SL}_n^1\big(\mathbb{F}_p[\![T]\!]\big)$ to prove the following:

PROPOSITION 12. *Suppose that the R-standard group G is R-perfect*

(1) *$[G_m, G_n] = G_{m+n}$ for all $n, m \in \mathbb{N}$.*
(2) *The lower central series and lower p-series coincide with the subgroups G_n*

$$G_n = \gamma_n(G) = P_n(G).$$

(3) *G is finitely generated and*

$$d(G) = d\big(G/\Phi(G)\big) = d\big(G/[G,G]G^p\big) \\ = \dim_{\mathbb{F}_p}(G/G_2) = d \cdot \dim_{\mathbb{F}_p}(\mathfrak{m}/\mathfrak{m}^2).$$

Given that R-perfect standard groups have the beginnings of a well-behaved theory perhaps we might have some hope to tackle the following:

PROJECT. (1) *Try to characterize group theoretically the fact that a pro-p group contains an R-perfect standard group.*
(2) *Determine whether R-perfect standard groups are linear.*

Bearing in mind that we achieved a diverse range of characterizations of p-adic analytic groups in terms of such properties as subgroup growth and the rank of a group, what prospect do these invariants offer for an approach to part (1) of our project?

Let us consider what the subgroup growth of such groups looks like. Recall that we defined

$$a_n(G) = \operatorname{card}\{H \leq G : |G{:}H| = n\}.$$

We recorded earlier that the polynomial growth of $a_n(G)$ for a pro-p group G was equivalent to the group having a p-adic analytic structure. Shalev [**32**] showed that there is a gap in the spectrum of subgroup growth of pro-p groups above polynomial:

THEOREM 8. *Let G be a pro-p group which is not p-adic analytic. Then there exists some constant $c \geq 1/8$ such that*

$$a_n(G) \geq n^{c \log_p n}$$

for infinitely many n.

Lubotzky and Shalev [**26**] used the graded Lie algebra $L(G)$ to prove that the R-perfect standard groups sit at the bottom of the spectrum of pro-p groups of non-polynomial subgroup growth:

THEOREM 9. *Let G be an R-perfect standard group. Then for a fixed constant c we have*

$$a_n(G) \leq n^{c \log_p n}$$

for all $n \geq 1$.

Shalev [**32**] proved that for example if $G = \mathrm{SL}_2^1(\mathbb{F}_p[\![T]\!])$ and $p > 2$ then

$$a_n(G) \leq 2n^{2 \log_p n}.$$

Is it possible that this type of growth characterizes R-perfect standard groups giving an answer to part (1) of our project?

To provide some answer to this question we return to the Nottingham group $\mathrm{Nott}(\mathbb{F}_p)$. It is clear that $\mathrm{Nott}(\mathbb{F}_p)$ is not an $\mathbb{F}_p[\![T]\!]$-perfect standard group since the shape of its lower central series as recorded in proposition 8 is not compatible with the uniform shape of the lower central series of an $\mathbb{F}_p[\![T]\!]$-perfect standard group. It is generally considered that $\mathrm{Nott}(\mathbb{F}_p)$ is not commensurable with any $\mathbb{F}_p[\![T]\!]$-standard group, i.e. that it is not an $\mathbb{F}_p[\![T]\!]$-analytic group. However this is still an open question:

PROBLEM. Prove that $\mathrm{Nott}(\mathbb{F}_p)$ is not an $\mathbb{F}_p[\![T]\!]$-analytic group.

Part of the evidence for this conviction comes from the fact that $\mathrm{Nott}(\mathbb{F}_p)$ is not linear since it contains every finitely generated pro-p group whilst all examples we know of $\mathbb{F}_p[\![T]\!]$-analytic groups are linear over $\mathbb{F}_p((T))$. However apart from this evidence, the Nottingham group $\mathrm{Nott}(\mathbb{F}_p)$ shares a remarkable amount in common with $\mathbb{F}_p[\![T]\!]$-perfect standard groups to the extent that any attempt to characterize $\mathbb{F}_p[\![T]\!]$-perfect standard groups in general catches the Nottingham group in its net. Take for example our proposal that perhaps R-perfect standard groups might be characterized by their subgroup growth. As the following Proposition reveals, the Nottingham group shares the same subgroup growth:

PROPOSITION 13. *If $p \geq 5$ then there exists a fixed constant $c(p)$ such that*

$$a_n(\mathrm{Nott}(\mathbb{F}_p)) \leq c(p) n^{(1+2/(p-1)) \log_p n}$$

for all $n \geq 1$.

An alternative characterization which also catches the Nottingham group is in terms of the width of a group. There seem to be two alternative definitions of width knocking around the literature. To keep things clear we shall rename one of these:

DEFINITION 22. Let G be a pro-p group

(1) The *lower-central width* of G is the supremum of the order of sections γ_i/γ_{i+1} in the lower central series γ_i of G.
(2) The *width* of G is the supremum of the order of all central sections of G.

A group of finite width obviously has finite lower central width. The definition of lower-central width generalizes the concept of the coclass of a pro-p group, a subject we have not touched on in this paper but an important chapter in the theory of pro-p groups. For references we refer the reader to Shalev's survey article [**33**] and the references contained there. We can add two further concepts of width to the above definitions by insisting only that the rank rather than the order of central sections be bounded. This concept of width generalizes the concept of the rank of a pro-p group and perhaps should be called *lower-central rand* and *central rank.*

In either event, $\mathbb{F}_p[\![T]\!]$-perfect standard groups and $\mathrm{Nott}(\mathbb{F}_p)$ both have finite width and hence finite lower-central width. So again these groups are found in the same net.

PROJECT. (1) *Classify thc just-infinite pro-p groups of finite lower-central width.*
(2) *Classify the just-infinite pro-p groups of finite width.*

Here just-infinite is the nearest we can get to a pro-p group being simple, namely that it should contain no infinite normal subgroups. The following answer was proposed to (1): either the group G is commensurable with a soluble group, or the positive part of a Loop group or the Nottingham group. However as we indicated in Example 11, work of Rozhkov proves that the pro-p completion of certain subgroups of the automorphism groups of p-regular trees also provide examples of groups of finite lower-central width.

There is another rather intriguing reason to view $\mathbb{F}_p[\![T]\!]$-perfect groups and $\mathrm{Nott}(\mathbb{F}_p)$ in the same camp. $\mathrm{Nott}(\mathbb{F}_p)$ also has a graded Lie algebra which is related to one of the non-classical characteristic p simple Lie algebras of Cartan type. Graded Lie algebras can be defined for any group G with a filtration G_n satisfying $[G_n, G_m] \leq G_{n+m}$. As for standard groups we define $L(G) = \bigoplus_{n\geq 1} G_n/G_{n+1}$. It turns out that $L(\mathrm{Nott}(\mathbb{F}_p))$ is isomorphic to the loop algebra of the first Witt Lie algebra W over $\mathbb{F}_p$. W is a simple Lie algebra with a basis $e_0, \dots, e_{p-1}$ satisfying

$$[e_i, e_j] = (j - i)e_{i+j \mod p}$$

W has a natural graded structure $W = \bigoplus_{i=0}^{p-1} W_i$ where $W_i = \langle e_i \rangle$. Then the graded Lie algebra $L(\mathrm{Nott}(\mathbb{F}_p))$ is the positive part L^+W of the loop algebra LW

defined by

$$L^{+}W = \bigoplus_{n \geq 1} W_n \mod p \otimes T^n.$$

We conclude this tour of the world of pro-p groups by gathering the characters we have met into a family photograph. At the bottom of the picture we have the old stalwarts the p-adic analytic pro-p groups like $\mathrm{SL}_n(\mathbb{Z}_p)$. Sitting above these are the less familiar pro-p groups with an analytic structure over more general pro-p rings, like $\mathrm{SL}_n(\mathbb{F}_p[\![T]\!])$ and $\mathrm{SL}_n(\mathbb{Z}_p[\![T]\!])$. The Nottingham group also likes to sit close to these groups as it has much in common with them. At the other end of the spectrum we have the free groups and some of the Galois groups we discussed. Other groups like the pro-p fractal groups (for example the pro-p completion of the Gupta-Sidki or Grigorchuk groups) have still to make up their mind where they are to sit and who knows, maybe the group you are interested in has a pro-p analogue which will provide a new face in this family photograph.

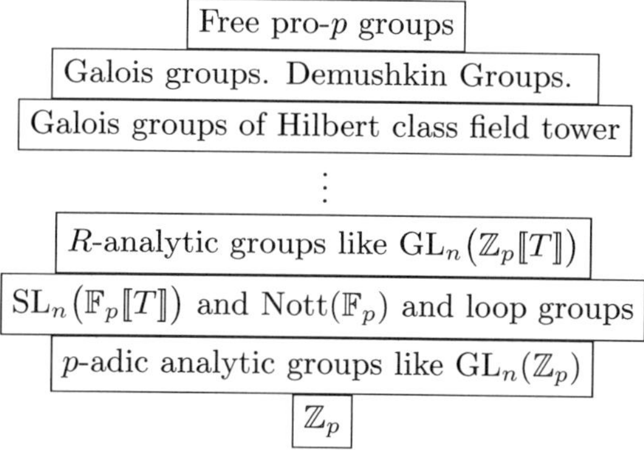

References

1. S. V. Aleshin, *Finite automata and Burnside's problem on periodic groups*, Mat. Zametki **11** (1972), 319–328.
2. N. Boston, *Some cases of the Fontaine-Mazur conjecture*, J. Number Theory **42** (1992), no. 3, 285–291.
3. ———, *Some cases of the Fontaine-Mazur conjecture.* II., preprint.
4. N. Bourbaki, *Commutative algebra*, Hermann and Addison-Wesley, 1972.
5. ———, *Lie groups and Lie algebras*, Hermann and Addison-Wesley, 1975.
6. R. Camina, *Subgroups of the Nottingham group*, J. Algebra **196** (1997), no. 1, 101–113.
7. D. Cartwright, *pro-p groups and Lie algebras over local fields of characteristic p*, Ph.D. thesis, Queen Mary and Westfield College, University of London.
8. J. D. Dixon, M. P. F. du Sautoy, A. Mann, and D. Segal, *Analytic pro-p groups*, London Math. Soc. Lecture Note Ser., vol. 157, Cambridge University Press, Cambridge, 1991.
9. M. P. F. du Sautoy, *Applications of p-adic methods to group theory*, p-adic Methods and their Applications (A. Baker and R. Plymen, eds.), Oxford Sci. Publ., Oxford Univ. Press, New York, 1992.
10. M. P. F. du Sautoy, A. Mann, and D. Segal, *Analytic pro-p groups* (updated edition in preparation).
11. J.-M. Fontaine and B. Mazur, *Geometric Galois representations*, Elliptic Curves and Modular Forms and Fermat's Last Theorem (Hong Kong, 1993), (J. H. Coates and S. T. Yau, eds.), Ser. Number Theory, I., Internat. Press, Cambridge, MA, 1995, 41–78.
12. E. Golod and I. Shafarevich, *On class field towers*, Izv. Akad. Nauk. SSSR **28** (1964), no. 1, 61–272. (Russian)
13. R. I. Grigorchuk, *On the Burnside problem for periodic groups*, Functional Anal. Appl. **14** (1980), 53–54.

14. ______, *On the Milnor problem of group growth*, Dokl. Akad. Nauk SSSR **271** (1983), no. 1, 30–33. (Russian)
15. N. Gupta and S. Sidki, *On the Burnside problem for periodic groups*, Math Z. **182** (1983), 385–388.
16. V. G. Kac, *Infinite dimensional Lie algebras*, 3rd ed., Cambridge University Press, Cambridge, 1990.
17. H. Koch, *Satz von Golod-Shafarevich*, Math. Nachr. **42** (1969), 321–333.
18. J. P. Labute, *Classification of Demushkin groups*, Canad. J. Math. **19** (1967), 106–132.
19. M. Lazard, *Groupes analytiques p-adiques*, Inst. Hautes Études Sci. Publ. Math. **26** (1965), 389–603.
20. A. Lubotzky, *Combinatorial group theory for pro-p groups*, J. Pure Appl. Algebra **25** (1982), 311–325.
21. ______, *Groups presentations, p-adic analytic groups and lattices in* $\mathrm{SL}_2(\mathbb{C})$, Ann. of Math. **118** (1983), 115–130.
22. ______, *Subgroup growth and congruence subgroups*, Invent. Math. **119** (1995), 267–295.
23. ______, *Eigenvalues of the Laplacian, the first Betti number and the congruence subgroup problem*, Ann. of Math. **145** (1997), 441–452.
24. A. Lubotzky and A. Mann, *Powerful pro-p groups* I. *and* II., J. Algebra **105** (1987), 484–515.
25. ______, *On groups of polynomial subgroup growth*, Invent. Math. **104** (1991), 521–533.
26. A. Lubotzky and A. Shalev, *On some* Λ*-analytic pro-p groups*, Israel J. Math. **85** (1994), no. 1–3, 307–337.
27. A. Pressley and G. Segal, *Loop groups*, Oxford Mathematical Monographs, Oxford Sci. Publ., The Clarendon Press, Oxford University Press, New York, 1986.
28. A. V. Rozhkov, *On the theory of groups of Aleshin type*, Mat. Zametki **40** (1986), no. 5, 572–589, 697. (Russian)
29. J.-P. Serre, Cohomologie galoisienne, 5th ed., Lecture Notes in Mathematics, vol. 5, Springer-Verlag, Berlin, 1994.
30. ______, *Lie algebras and Lie groups*, Lecture Notes in Math., vol. 1500, Springer-Verlag, Berlin, 1992.
31. I. R. Shafarevich, On p-extensions , Math. Sbornik **20** (1947), 351–363 (Russian).
32. A. Shalev, *Growth functions, p-adic analytic groups, and groups of finite coclass*, J. London Math. Soc. (2) **46** (1992), 111–122.
33. ______, *Some problems and results in the theory of pro-p groups*, Groups '93 Galway/St. Andrews, vol. 2, London Math. Soc. Lecture Note Ser., vol. 212, Cambridge Univ. Press, Cambridge, 1995, pp. 528–542.
34. V. I. Sushchansky, *Periodic p-groups of permutations and the general Burnside problem*, Dokl. Akad. Nauk SSSR **247** (1979), 447–461.
35. T. S. Weigel, *On the profinite completion of arithmetic groups of split type*, preprint.

Department of Pure Mathematics and Mathematical Statistics, 16 Mill Lane, Cambridge CB2 1SB, U.K.

E-mail address: dusautoy@dpmms.cam.ac.uk

Centre de Recherches Mathématiques
CRM Proceedings and Lecture Notes
Volume **17**, 1999

Identities of Representations of Finite Groups

Samuel M. Vovsi

Introduction

Let F be the free group of countable rank on an alphabet $x_1, x_2, \ldots$, and let KF be its group algebra over a field K. An element $u = u(x_1, \ldots, x_n) \in KF$ is called an *identity* of a representation $\rho\colon G \to \mathrm{GL}_K(V)$ if for arbitrary $g_1, \ldots, g_n \in G$ the element $u(g_1, \ldots, g_n) \in KG$ annihilates V.

The present paper is concerned with identities of representations of finite groups and with some related questions on varieties of group representations, i.e. classes of group representations definable by identities. It is intended for the non-expert and provides a brief survey of some aspects of the field. To put it in perspective, recall that a systematic study of varieties of group representations was initiated in the early seventies in a series of papers by B. I. Plotkin and his students. The principal original contribution of that work was that for the first time the idea of variety was applied to algebraic structures with *two underlying sets*. Very soon it became clear that identities offer a fruitful new viewpoint in representation theory, and that the theories of varieties of representations of groups, Lie algebras, semigroups, etc. are of interest in their own right. At the present time, there is no doubt that the study of identities of algebraic structures is closely connected with the study of identities of their linear representations. This point of view was systematically exploited in the books [**PV**] and [**R2**] and found applications in a number of papers. The first applications of this sort were demonstrated by Bryce [**Br**] and Ol'shansky [**O**], who used the methods of varieties of group representations in the study of varieties of abstract group. Several principal results of Razmyslov (see for example [**R1**]) were essentially based on his technique of weak identities, i.e. identities of representations of Lie algebras. In papers of Krasil'nikov [**K1, K2**] and the author [**V2**], methods of varieties of group representations were applied to the study of polynomial identities of associative algebras. Finally, one should mention an important work of Il'tyakov [**I**]. Its main result asserts that every PI-representation of a finitely generated Lie algebra over a field of characteristic 0 has a finite basis of identities. One corollary of this result is the finite-dimensional case of Kemer's theorem. Another major corollary is that in characteristic zero every finite-dimensional Lie algebra has a finite basis of identities.

1991 *Mathematics Subject Classification.* Primary: 20E10; Secondary: 20C99.
This is the final form of the paper.

The paper consists of two sections which deal with two different aspects of the same theme: the finite basis problem. In Section 1, the main question is whether all identities of an arbitrary representation of a finite group are consequences of some finite subset of identities. In other words, does there exist a finite basis for the identities of every such representation? We describe results obtained in this direction and discuss a few related open problems. Since these results are well known, proofs are generally omitted. The only exception is the proof of the fact that an ordinary representation of a finite group has a finite basis of identities (Proposition 1). We repeat this proof here in order to give an idea of the techniques used, and because it is very short and simple.

Section 2 is a natural continuation of Section 1. However, while the latter is concerned with the *pure existence* of a finite basis of identities, in the former we are interested in computing such a basis *explicitly*. More specifically, we are interested in finding a basis for the identities of the ordinary regular representation of a given finite group G. The technique used for this purpose might be of interest both from the standpoint of model theory and finite group theory. It was suggested in [**PK**] (see also [**PV**]; Section 13) and reduces the problem to describing the *disjunctive identities* of G, i.e. certain universal formulas of first order logic which are identically true on G. We develop this method further and apply it to compute bases for the identities of certain representations.

It should be emphasized that the main goal of Section 2 was not these bases themselves but the *process*. We attempted to streamline the method of [**PK**] and develop an almost mechanical procedure which would allow to find such bases in a finite number of elementary steps for many different groups. To demonstrate how this procedure works, we applied it to two simple examples (the groups D_8 and A_5), deliberately trying to keep our exposition elementary. These examples show that, in principle, for groups of small order (say, ≤ 100) this work can be done by a good undergraduate student. However, the complexity of the work increases very rapidly with the order of the group in question. It would be very interesting to understand whether a computer can be involved in this process.

I am grateful to Gregory Cherlin for useful conversations and comments on the first draft of this paper.

ADDED IN PROOF. While this paper was in print, a detailed exposition of the material of Section 2 appeared in [**V5**].

1. Existence of a Finite Basis

First we recall a few basic definitions and notation. For the details the reader is referred to [**PV**] or [**V3**].

Let K be an arbitrary but fixed field. A representation $\rho\colon G \to \mathrm{GL}_K(V)$ of a group G on a K-space V is often denoted by $\rho = (V, G)$; the result of the action of $g \in G$ on $v \in V$ is denoted by vg. A class of group representations is called a *variety* if it consists of all representations satisfying a certain set of identities. If $\mathcal{X}$ is a variety of group representations over K, then the set $\operatorname{Id}\mathcal{X}$ of all its identities is a two-sided ideal of KF, which is invariant under all endomorphisms of the group F. Such ideals are called *fully invariant* (or *verbal*). A standard argument shows that the map $\mathcal{X} \mapsto \operatorname{Id}\mathcal{X}$ is a bijection between the varieties and the fully invariant ideals of KF.

A variety $\mathcal{X}$ is called *finitely based* if it can be defined by a finite number of identities. This is equivalent to saying that $\operatorname{Id}\mathcal{X}$ is finitely generated as a fully invariant ideal of KF. A representation ρ is finitely based if it generates a finitely based variety.

EXAMPLE.

(1) A representation $\rho = (V, G)$ is called *stable of class n*, or simply *n-stable* (this terminology goes back to Kaloujnine and P. Hall) if there is a series of G-modules

$$0 = A_0 \subseteq A_1 \subseteq \cdots \subseteq A_n = V$$

such that G acts trivially on each quotient A_{i+1}/A_i. A typical example of an n-stable representation is $ut_n(K) = \big(K^n, UT_n(K)\big)$, where K^n is the n-dimensional space and $UT_n(K)$ is the full unitriangular matrix group of degree n acting on K^n in the natural way. The class $\mathcal{S}^n$ of all n-stable representations is a finitely based variety because it is definable by a single identity $(x_1 - 1)(x_2 - 1)\cdots(x_n - 1)$. It is easy to see that $\operatorname{Id}(\mathcal{S}^n) = \Delta^n$, where Δ is the augmentation ideal of KF.

(2) Another well known example of a finitely based variety is the variety $\mathcal{U}_n$ of all *n-unipotent* representations, defined by the identity $(x-1)^n$. Evidently $\mathcal{S}^n \subseteq \mathcal{U}_n$. On the other hand, by a classical theorem of Kolchin, every finite-dimensional unipotent representation is stable. In other words, for finite-dimensional representations "stable" and "unipotent" are the same. It can be deduced from Zelmanov's theorem [**Z**] on the nilpotency of Engel Lie algebras that over a field of characteristic 0 this is true even for infinite-dimensional representations. Over a field of prime characteristic it is not the case.

(3) For any variety of groups $\mathfrak{V}$, denote by $\omega\mathfrak{V}$ the class of all representations $\rho = (V, G)$ such that $G/\operatorname{Ker}\rho \in \mathfrak{V}$. This class is a variety because if $\mathfrak{V}$ is defined by group words f_i, then $\omega\mathfrak{V}$ is defined by the identities $f_i - 1$. It is easy to see that the map $\mathfrak{V} \mapsto \omega\mathfrak{V}$ is injective and that $\omega\mathfrak{V}$ is finitely based if and only if $\mathfrak{V}$ is finitely based. The ideal $\operatorname{Id}(\omega\mathfrak{V})$ of identities of $\omega\mathfrak{V}$ is generated, as a right ideal, by all $f - 1$, where f belongs to the $\mathfrak{V}$-verbal subgroup $\mathfrak{V}(F)$. Equivalently, $\operatorname{Id}(\omega\mathfrak{V})$ is the kernel of the natural epimorphism $KF \to K\big[F/\mathfrak{V}(F)\big]$

One of the first questions that naturally arise in the theory of varieties of arbitrary algebraic structures is the following: does every finite object have a finite basis of identities? Without going into details, we recall that this was solved in the affirmative for groups (Oates–Powell [**OP**]), associative rings (Kruse [**Kr**] and L'vov [**L**]), Lie rings (Bakhturin–Ol'shanskii [**BO**]), lattices (McKenzie [**Mc**]), and many other algebras. On the other hand, there are examples of finite algebras (e.g. semigroups) having no finite bases for their identities.

Does every representation of a finite group have a finite basis of identities? This question has attracted attention since the early stages of the theory of varieties of group representations. The first partial result was obtained by the author [**V1**] who proved that for ordinary (i.e. nonmodular) representations the answer is positive. It is a very simple fact and its proof will be given below. Later this result was generalized in [**P**], where it was proved that there is a finite basis of identities for every representation $\rho\colon G \to \mathrm{GL}_K(V)$ satisfying the following three conditions:

(a) V is finite-dimensional;

(b) G has a normal subgroup H of finite index acting stably on V;
(c) $|G/H|$ is not divisible by the characteristic of the ground field.

At the same time, the initial question on the existence of a finite basis for an arbitrary representation of a finite group remained open for a rather long time. Moreover, the results of [**V1**] and [**P**] led to a conjecture that every *stable-by-finite* representation (i.e. representation satisfying only the condition (b)) has a finite basis of identities. This conjecture was eventually proved in [**VN**].

THEOREM 1 [**VN**]. *The variety generated by an arbitrary stable-by-finite representation is a variety of bounded rank, that is*
(a) *it is finitely based*;
(b) *it is locally finite-dimensional*;
(c) *all its subvarieties have finite and uniformly bounded basis ranks.*

The next natural step in this direction is to consider the finite basis problem for representations over an arbitrary commutative ring K. As usual in such situations, K must be noetherian because otherwise one can easily indicate trivial counterexamples.[1] This leads to

PROBLEM 1. Does every representation of a finite group over a noetherian commutative ring K have a finite basis of identities? Is this true at least if $K = \mathbb{Z}$?

We were able to obtain only a partial result on this problem.

THEOREM 2 [**V4**]. *Let K be a commutative noetherian ring and let $\rho: G \to \mathrm{GL}_K(V)$ be a representation of a finite group G on a K-module V. If either V is a module of finite length or $nV = 0$ for some positive integer n, then ρ has a finite basis of identities.*

In particular, suppose that K is artinian. Let $\rho: G \to \mathrm{GL}_K(V)$ be a representation of a finite group G over K, and let $\mathfrak{X} = \operatorname{var}\rho$ be the variety of group representations generated by ρ. An elementary observation shows that $\mathfrak{X}$ can be generated by a representation of G on some *finitely generated* K-module W. Since K is artinian, W is of finite length. Hence Theorem 2 implies

COROLLARY 1. *Every representation of a finite group over an artinian commutative ring has a finite basis of identities.*

To explain some of the remaining difficulties (say, for $K = \mathbb{Z}$), let us first consider the simplest case, which was settled long ago in [**V1**]. Namely, let us prove that every ordinary representation of a finite group is finitely based.

The ideology of the proof goes back to Oates and Powell [**OP**] and is based on two principal concept: critical representations and Cross varieties. Recall that a representation is *critical* if it does not belong to the variety generated by its proper sections. It is well known that a finite simple algebra is "usually" critical—this is known for groups, rings, Lie rings and, in general, for algebras from any congruence-modular variety. Similarly, a simple (i.e. faithful and irreducible) representation of a finite group is critical. A crucial observation is that for ordinary representations of finite groups the reverse is also true (an immediate consequence of the Maschke Theorem), and so in our situation

$$\text{critical} = \text{faithful} + \text{irreducible}.$$

[1] Alternatively, the problem can be formulated over an arbitrary K, but then one should impose certain restrictions on the identities in question.

A variety $\mathcal{X}$ of group representations is called *locally finite* if $\mathcal{X} \subset \omega\mathfrak{V}$ for some locally finite variety of groups $\mathfrak{V}$. If $\mathfrak{V}$ can be chosen with exponent not divisible by the characteristic of the ground field, then $\mathcal{X}$ is said to be *ordinary*. $\mathcal{X}$ is called *Cross* if

(i) it is finitely based;
(ii) it is locally finite;
(iii) it contains only a finite number of nonisomorphic critical representations.

It is well known that a subvariety of a Cross variety is also Cross and that every locally finite variety is generated by its critical representations (cf. [**N**]; Section 5.1).

PROPOSITION 1 [**V1**]. *Let $\rho: G \to \mathrm{GL}_K(V)$ be an ordinary representation of a finite group G. Then the variety generated by ρ is Cross.*

We will need one intermediate fact. In the free group algebra KF, define inductively

$$w_1 = (x_0^{-1}x_1)^{x_{01}} - 1,$$
$$w_d = w_{d-1}\big((x_0^{-1}x_d)^{x_{0d}} - 1\big)\dots\big((x_{d-1}^{-1}x_d)^{x_{d-1,d}} - 1\big),$$

where x_i, x_{ij} are pairwise distinct free generators of F.

LEMMA 1. *Let $\rho = (V, G)$ be an arbitrary representation. If $|G| \le d$, then w_d is an identity of ρ. Conversely, if w_d is an identity of ρ and ρ is faithful and irreducible, then $|G| \le d$.*

PROOF. The first claim is obvious. To prove the second, assume that ρ is faithful and irreducible. Note that if $0 \ne v \in V$ and $1 \ne g \in G$, then there exists $h \in G$ such that $v(g^h - 1) \ne 0$. Indeed, let W be the G-submodule of V generated by all vh^{-1}, $h \in G$. Since V is irreducible, $W = V$. Since $g \ne 1$, there is vh^{-1} such that $(vh^{-1})g \ne vh^{-1}$, or $v(g^h - 1) \ne 0$.

Assume now that w_d is valid in ρ and that $|G| > d$. Choose $d+1$ distinct elements $g_1, g_2, \dots, g_{d+1} \in G$ and let $0 \ne v \in V$. Since $g_0^{-1}g_1 \ne 1$, there exists $h_{01} \in H$, such that

$$v_1 = v\big((g_0^{-1}g_1)^{h_{01}} - 1\big) \ne 0.$$

Since $g_0^{-1}g_2 \ne 1$, $g_1^{-1}g_2 \ne 1$, etc., we can find $h_{02}, h_{12}, \dots \in G$, such that

$$v_2 = v_1\big((g_0^{-1}g_2)^{h_{02}} - 1\big) = v\big((g_0^{-1}g_1)^{h_{01}} - 1\big)\big((g_0^{-1}g_2)^{h_{02}} - 1\big) \ne 0,$$
$$v_3 = v_2\big((g_1^{-1}g_2)^{h_{12}} - 1\big) = v \cdot w_2(g_1, g_2, h_{01}, h_{02}, h_{12}) \ne 0$$

and so on. Repeating this process, we will eventually see that $v \cdot w_d \equiv 0$ fails in ρ, which is a contradiction.

PROOF OF PROPOSITION 1. Let $\mathfrak{V} = \operatorname{var} G$ be the variety of groups generated by G. Since, by the Oates-Powell theorem, $\mathfrak{V}$ is finitely based, $\omega\mathfrak{V}$ is finitely based as well. Let $|G| = d$ and let $\mathcal{X}$ be the variety consisting of all representations from $\omega\mathfrak{V}$ satisfying the identity w_d. Then $\rho \in \mathcal{X}$ and so $\operatorname{var}\rho$ is a subvariety of $\mathcal{X}$. Therefore it suffices to show that $\mathcal{X}$ is Cross. Since $\mathcal{X} \subseteq \omega\mathfrak{V}$, it is locally finite. Since $\omega\mathfrak{V}$ is finitely based, so is $\mathcal{X}$. Let $\sigma = (W, H)$ be a critical representation in $\mathcal{X}$. By assumption, ρ is an ordinary representation, that is, $\operatorname{char} K \nmid |G|$. It follows that the exponent of the group variety $\mathfrak{V}$ is not divisible by $\operatorname{char} K$ either. Therefore the representation $\sigma = (W, H)$ is also ordinary. Since ordinary critical representations are faithful and irreducible and σ satisfies w_d, Lemma 1 implies that $|H| \leq d$. Thus $\mathcal{X}$ has only a finite number of critical representations and so $\mathcal{X}$ is Cross.

This proof turned out to be so easy because critical *ordinary* representations have an extremely simple characterization: they are just faithful irreducible representations of finite groups. But critical *modular* representations need not be irreducible, and there are examples showing that a variety generated by a modular representation of a finite group need not be Cross. Still, by Theorem 1, it is a variety of bounded rank and therefore is finitely based and satisfies both the maximum and the minimum conditions on subvarieties (see [**VN**] or [[**V3**]; Chapter 3] for the details). The proof in the modular case is much more difficult and is based on a careful analysis of the critical objects in the coresponding variety.

If K is not a field, the situation is even more complicated. Consider a simple example. Let $K = \mathbb{Z}, G$ a finite p-group, and let $\mathrm{Reg}_{\mathbb{Z}} G = (\mathbb{Z}G, G)$ be the regular representation of G. For each $k = 1, 2, \ldots,$ the regular representation $\rho_k = (\mathbb{Z}_{p^k}G, G)$, regarded as a representation over $\mathbb{Z}$, is a homomorphic image of $\mathrm{Reg}_{\mathbb{Z}} G$ and hence belongs to $\mathrm{var}(\mathrm{Reg}_{\mathbb{Z}} G)$. All the ρ_k are stable (= unipotent) and their indices of stability are unbounded. It follows that $\mathrm{var}(\mathrm{Reg}_{\mathbb{Z}} G)$ does not satisfy the maximum condition on subvarieties. This shows that there may be serious barriers to the solution of our problem, and a possibility of a negative answer is not excluded. The crucial point will be to understand the structure of critical representations in the corresponding variety.

Even if K is a field, the structure of critical (modular!) representations is by no means well understood. Since every locally finite variety is generated by its critical representations, the study of these objects may lead to a deeper understanding of a field as a whole. In our opinion, one problem on critical representations is especially interesting. Let $\rho: G \to \mathrm{GL}_K(V)$ be a representation. The G-module V is called *monolithic* (*comonolithic*) if it has a smallest (largest) nontrivial G-submodule; the same terms apply to the representation ρ.

LEMMA 2. *A critical representation is both monolithic and comonolithic.*

PROOF. Let us prove, for example, the second claim. Let $\rho = (V, G)$ be a representation. If it is not comonolithic, then $V = V_1 + V_2 + \ldots,$ where the V_i are proper G-submodules of V. Consider the direct sum of G-modules $V' = V_1 \oplus V_2 \oplus \ldots$ and the corresponding representation $\rho' = (V', G)$. Then ρ' belongs to the variety generated by the proper subrepresentations $\rho_i = (V_i, G)$ of ρ. Since there is a natural epimorphism of G-modules $V' \to V$, so does ρ. It follows that ρ is not critical.

Now let G be a finite group and let

$$KG = M_1 \oplus M_2 \oplus \cdots \oplus M_n$$

be a decomposition of the group algebra KG into a direct sum of indecomposable G-modules. These modules, known as the *principal indecomposables* of G, play a fundamental role in modular representation theory. It is well known that they are simultaneously monolithic and comonolithic. Together with Lemma 2, this suggest

PROBLEM 2. Are the representations of a finite group G corresponding to its principal indecomposables critical? In particular, is this true if $G = S_n$?

To conclude this section, we note that it is still unknown whether every finite-dimensional representation over a field has a finite basis of identities.[2] Undoubtedly, this is one of the major problems of the theory of varieties of group representations, and up to now it remains out of reach. But the following important particular case seems to be more accessible.

PROBLEM 3. Is it true that every finite-dimensional representation of a solvable group has a finite basis of identities?

Our optimism is based on two observations. Let $\rho: G \to \mathrm{GL}_K V$ be a finite-dimensional representation of a solvable group G. Without loss of generality we may assume ρ to be faithful, and so G is a solvable matrix group. By the classical Kolchin–Mal'tsev Theorem, G has a normal series

$$N \lhd H \lhd G$$

such that N is a unitriangular matrix group (and so N acts stably on V), H/N is abelian and G/H is finite.

OBSERVATION 1. Suppose the section H/N is trivial. Then ρ is stable-by-finite and, by Theorem 1, is finitely based.

OBSERVATION 2. Suppose that G/H is trivial. Then ρ is stable-by-abelian and, as proved by Krasil'nikov [**K1, K2**], is finitely based.

One might expect that a complete solution of Problem 3 can be obtained by a suitable combination of methods from [**VN**] and [**K1**]. Unfortunately (or fortunately) the methods of these papers are radically different, and it will not be easy to make them work together.

2. Computation of a Finite Basis

Let $\rho: G \to \mathrm{GL}_K V$ be a representation of a finite group G over a field K. In §1 we discussed the problem of *existence* of a finite basis of identities for ρ. In the present section we are interested in finding such a basis *explicitly*. Of course, there is no hope to develop a universal method allowing to write down a basis of identities for an arbitrary representation, so we will be concerned with a much more concrete (but still very general) problem.

PROBLEM 4. Let G be a finite group, $\operatorname{char} K \nmid |G|$. Find a basis of identities for the regular representation $\mathrm{Reg}_K G = (KG, G)$ of G over K.

Why are the regular representations of particular interest? The reason is two-fold. First, it is easy to understand that the identities of $\mathrm{Reg}_K G$ are the identities of *all* representations of G over K. Second, they can be regarded as special "weak" identities of the group algebra KG, for they are exactly the elements $u(x_1, \dots, x_n) \in KF$ such that $u(g_1, \dots, g_n) = 0$ in KG for all $g_i \in G$.

An interesting approach to Problem 4 was suggested by Plotkin and Kushkuley [**PK**]. Consider a universally quantified formula

$$\delta = \forall x_1 \dots x_n \big[(f_1 = 1) \vee (f_2 = 1) \vee \cdots \vee (f_n = 1)\big] \tag{1}$$

[2] Here the acting group need not be finite.

where $x_1, \ldots, x_n$ are all variables involved in the f_i. The first systematic study of such formulas was probably done by Baker [**B**], who called them universal disjunctions of equations (UDE). We say that δ is a *disjunctive identity* (d-identity, for short) of a group G if it is identically true on G. It is obvious that if δ is a d-identity of G then

$$\delta^* = (f_1^{y_1} - 1) \ldots (f_n^{y_n} - 1) \in KF$$

is an identity of every representation of G. It turns out that if one knows the d-identities of a finite group G, then one can easily obtain a basis of identities of the regular representation $\mathrm{Reg}_K G$. To make this more precise, we recall a few simple definitions and facts.

Let δ be an arbitrary d-identity, that is, a universal sentence of the form (1). Usually we omit the quantor prefix and write this formula simply as $\dot{=}(f_1 = 1) \vee (f_2 = 1) \vee \cdots \vee (f_n = 1)$. A class of groups $\mathfrak{D}$ is called a *d-variety* if it is definable by a set of d-identities.[3] This is equivalent to saying that $\mathfrak{D}$ is definable by positive universal sentences, because every such sentence is obviously equivalent to a conjunction of d-identities. Applying a well known fact of model theory ([**G**] p. 275), we see that a class of groups is a d-variety if and only if it is closed under taking subgroups, homomorphic images, and ultraproducts.

If $\mathfrak{X}$ is an arbitrary class of groups, then by $\mathrm{dvar}(\mathfrak{X})$ we denote the d-variety generated by $\mathfrak{X}$. In other words, $\mathrm{dvar}(\mathfrak{X})$ is the class of all groups satisfying all d-identities that are valid in all groups from $\mathfrak{X}$. Let G be a finite group and let $\mathrm{H}S(G)$ be the class of all sections of G. Being a finite class of finite algebras, $\mathrm{H}S(G)$ is axiomatizable by a single sentence. Therefore it is closed under ultraproducts and so is a d-variety. Thus we have proved

LEMMA 3. *The d-variety generated by a finite group is the class of all sections of this group.*

In particular, this implies that if G and H are finite groups then

$$\mathrm{dvar}(G) = \mathrm{dvar}(H) \iff G \simeq H.$$

In other words, *every finite group is uniquely determined by its d-identities.*

A set of d-identities $\{\delta_i\}$ of a class of groups $\mathfrak{X}$ is called a *basis* of d-identities of $\mathfrak{X}$ if every group satisfying the δ_i belongs to $\mathrm{dvar}(\mathfrak{X})$ (that is, satisfies *all d-identities of* $\mathfrak{X}$). It is easy to see that every finite group has a finite basis of d-identities. Indeed, let G be group of order n. Consider the d-identity

$$\omega_n = \bigvee_{0 \le i < j \le n} (x_i = x_j)$$

and for every group H_i of order at most n and not belonging to $\mathrm{dvar}(G)$ choose a d-identity δ_i of G which is not valid in H_i. Then it is clear that a finite set of formulas $\{\omega_n, \delta_1, \delta_2, \ldots\}$ is a basis of d-identities of G, as required.[4]

We now discuss the connection between d-identities of groups and identities of their regular representations. Let G be a group and let $\{\delta_i\}$ be a set of d-identities

[3]Initially, Plotkin and Kushkuley [**PK**] used the words "pseudoidentity" and "pseudovariety". We would also prefer to use this (pseudo)terminology but, unfortunately, it has already been occupied by Eilenberg and Schützenberger [**ES**].

[4]All of the above is based on general results of model theory and is valid not only for groups, but for arbitrary algebraic systems.

of G. This set is called a *weak basis* of d-identities of G if every irreducible linear (over K) group satisfying all the δ_i belongs to dvar(G) (i.e. is a section of G). Of course, every basis of d-identities is a weak basis, but the converse, in general, is not true. The following fact plays a key role.

LEMMA 4 [**PK**]. *Let G be a finite group and let U be a set of identities of the regular representation* $\text{Reg}_K G = (KG, G)$, *satisfying the following conditions*:
(i) *the variety of representations defined by U is locally finite and ordinary*;
(ii) *there exists a weak basis $\{\delta_i\}$ of d-identities of G such that $U \supseteq \{\delta_i^*\}$.*
Then U is a basis of identities of $\text{Reg}_K G$.

PROOF. The proof of this statement is not difficult, and we will provide it here. Take $\delta = (f_1 - 1) \vee \cdots \vee (f_n - 1)$, and let $\delta^* = (f_1^{y_1} - 1) \dots (f_n^{y_n} - 1)$. First we show that if δ^* is an identity of some faithful irreducible representation $\rho = (V, H)$, then δ must be a d-identity of H. Assume the contrary; then there is a homomorphism $\phi: F \to H$ such that $f_i^\phi \neq 1$ for every i. Denote $f_i^\phi = g_i$ and take $0 \neq v \in V$. Now we use the same argument as in the proof of Lemma 1. Since V is an irreducible KH-module, we have $vKH = V$. Since H acts on V faithfully and $g_1 \neq 1$, there exists $h_1 \in H$ such that $v(g_1^{h_1} - 1) \neq 0$. Next, denote $v(g_1^{h_1} - 1) = v_1$ and find $h_2 \in H_2$ such that $v_2 = v_1(g_2^{h_2} - 1) \neq 0$. Repeating this argument, we find elements $h_3, \dots, h_n \in H$ such that

$$v \cdot (g_1^{h_1} - 1)(g_2^{h_2} - 1) \dots (g_n^{h_n} - 1) \neq 0.$$

But this is impossible because δ^* is an identity of ρ.

Now let G and U be as in the statement. Let $\mathcal{X}$ be the variety generated by $\text{Reg}\, G$ and let $\mathcal{Y}$ be the variety defined by U. Then $\mathcal{X} \subseteq \mathcal{Y}$; we have to show that $\mathcal{X} = \mathcal{Y}$. By (i), $\mathcal{Y}$ is locally finite and ordinary, and therefore it is generated by faithful irreducible representations of finite groups. Take such a representation $\rho = (V, H)$ in $\mathcal{Y}$; it remains to show that $\rho \in \mathcal{X}$.

Since the δ_i^* belong to U, they all are identities of ρ. By the above, the δ_i are d-identities of H. Since H is an irreducible linear group and $\{\delta_i\}$ is a weak basis of d-identities of G, it follows that $H \in \text{dvar}(G)$, that is, H is a section of G. But then $\text{Reg}\, H$ satisfies all identities that are valid in $\text{Reg}\, G$. Hence $\text{Reg}\, H$ belongs to $\mathcal{X}$, and the same is true for *any* representation of H.

Since it is very easy to find identities of the representation $\text{Reg}_K G$ which guarantee that the corresponding variety is locally finite and ordinary, Lemma 4 reduces solution of Problem 4 to finding a weak basis of d-identities of a given finite group G. These ideas were developed in [**PK**] in several directions; in particular, they were applied to the study of identities of the regular representations of the symmetric groups S_n for $n \leq 5$.

The present section is focused on finding *actual* (not weak) bases of d-identities.

PROBLEM 5. For a given finite group G, find a basis of its d-identities.

This problem is more difficult than its "weak sibling" and is of much greater independent interest. Indeed, as mentioned earlier, every finite algebra is uniquely determined by its disjunctive identities, and therefore a basis of d-identities of a finite group uniquely identifies it. What is especially important, Problem 5 is in a certain sense "finite", and if the order of G is small enough, it can be solved by a rather mechanical procedure. We illustrate this by the following two examples.

EXAMPLE 1. Let D_8 be the dihedral group of order 8:

$$D_8 = \langle a, b \mid a^4 = b^2 = 1, a^b = a^{-1} \rangle.$$

Clearly it satisfies the d-identities

$$\omega_8 = \bigvee_{0 \le i < j \le 8} (x_i = x_j) \tag{2}$$

and

$$x^4 = 1. \tag{3}$$

Since D_8 contains only two elements of order > 2, it also satisfies the d-identity

$$(x_1^2 = 1) \vee (x_2^2 = 1) \vee (x_3^2 = 1) \vee (x_1 = x_2) \vee (x_1 = x_3) \vee (x_2 = x_3). \tag{4}$$

We prove that the formulas (2)–(4) form a *weak* basis of d-identities of D_8. We have to show that every irreducible linear group G satisfying (2)–(4) belongs to $\operatorname{dvar}(D_8)$, i.e. is a section of D_8. Since G satisfies ω_8, it must be of order at most 8. The groups of orders 3, 5, 6, 7 are excluded by (3), but the groups of orders 1, 2, 4 are all sections of D_8. There are five groups of order 8:

$$Z_8, Z_4 \times Z_2, Z_2 \times Z_2 \times Z_2, Q_8, \text{ and } D_8,$$

where, as usual, Z_n is the cyclic group of order n and Q_8 is the quaternion group. Z_8 does not satisfy (3), each of the groups $Z_4 \times Z_2$ and Q_8 has three distinct elements of order 4 and so does not satisfy (4), and Z_2^3 is not irreducible linear (an abelian irreducible linear group must be cyclic). This completes the proof.

However, the formulas (2)–(4) do not form a basis of d-identities of D_8 because Z_2^3 obviously satisfies all of them but is not a section of D_8. To eliminate Z_2^3, we have to introduce another d-identity. For brevity, for any variables x and y we will write $x \in \langle y \rangle$ to denote

$$(x = 1) \vee (x = y) \vee (x = y^2) \vee (x = y^3).$$

Consider the following formula in the variables x_1, x_2, x_3:

$$\begin{aligned} \Big[\bigvee_{i \neq j} x_i \in \langle x_j \rangle\Big] &\vee \Big[\bigvee_{\sigma \in S_3} x_{\sigma(1)} \in \langle x_{\sigma(2)} x_{\sigma(3)} \rangle\Big] \\ &\vee \Big[\bigvee_{\sigma \in S_3} \langle x_{\sigma(1)} \rangle \ni x_{\sigma(2)} x_{\sigma(3)}\Big] \vee \Big[\bigvee_{\sigma \in S_3} x_{\sigma(1)} x_{\sigma(2)} \in \langle x_{\sigma(2)} x_{\sigma(3)} \rangle\Big] \end{aligned} \tag{5}$$

and show that it is a d-identity of D_8. The group D_8 is a semidirect product

$$D_8 = A \leftthreetimes B, \text{ where } A \simeq Z_4,\ B \simeq Z_2.$$

Note that if x, y are elements of D_8 not in A, then $xy \in A$. Now take arbitrary x_1, x_2, $x_3 \in D_8$. If two of the x_i lie in A then the first clause of (5) holds, if one of them lies in A and the other two outside then the second or the third clause holds, and if all three x_i lie outside then the last clause holds.

On the other hand, (5) is not valid in Z_2^3: it is enough to take a basis of this group for x_1, x_2, x_3. Thus we have proved that the formulas (2)–(5) form a basis for the d-identities of D_8.

It is now trivial to find a basis for the identities of the regular representation of D_8, provided $\operatorname{char} K \neq 2$. We know that (2)–(4) is a weak basis of d-identities of D_8. Applying the "*-operation" to these three formulas, we obtain the following elements of KF:

$$\prod_{0\le i<j\le 8} \left((x_i x_j^{-1})^{y_{ij}} = 1\right), \tag{6}$$

$$x^4 - 1, \tag{7}$$

$$(x_1^{2y_1} - 1)(x_2^{2y_2} - 1)(x_3^{2y_3} - 1)\left((x_1x_2^{-1})^{y_4} - 1\right) \times \left((x_1x_3^{-1})^{y_5} - 1\right)\left((x_2x_3^{-1})^{y_6} - 1\right). \tag{8}$$

The variety of group representations defined by (6)–(8) is obviously contained in $\omega\mathfrak{B}_4$, where $\mathfrak{B}_4$ is the Burnside variety of exponent 4. Since $\omega\mathfrak{B}_4$ is locally finite and ordinary, Lemma 4 implies that (6)–(8) is a basis of identities of $\operatorname{Reg} D_8$.

EXAMPLE 2. Our next aim is to compute a basis for the d-identities of the smallest non-abelian simple group—the alternating group A_5—and a basis for the identities of $\operatorname{Reg} A_5$. We will often use the following obvious facts:

(a) if a set of d-identities eliminates all groups of order p^k with p a prime, then the same set eliminates all groups of order divisible by p^k (by Sylow's theorem);

(b) if a set of d-identities eliminates all groups of order m, then the same set eliminates all solvable groups of order mn with $(m,n) = 1$ (by Hall's theorem).

Also, we will sometimes informally say that the permutation (123)(45) has a cycle form $(*\,*\,*)\,(**)$, the permutation (15)(32) has form $(**)\,(**)$, etc.

1. Every nonunit element of A_5 has one of the following three cycle forms: $(*\,*\,*\,*\,*)$, $(*\,*\,*)$, $(**)\,(**)$. Therefore A_5 satisfies the d-identity

$$(x^2 = 1) \vee (x^3 = 1) \vee (x^5 = 1) \tag{9}$$

and, of course, ω_{60}. These two d-identities may both be valid in the groups of the following orders only:

$$1, 2, 3, 4, 5, 6, 8, 9, 10, 12, 15, 16, 18, 20, 24, 25, 27, 30, 32, 36, 40, 45, 48, 50, 54, 60.$$

The groups of order 1, 2, 3, 5 are all subgroups of A_5. The cyclic group of order 4 does not satisfy (9), but the noncyclic one is a subgroup of A_5. Similarly, the cyclic group of order 6 does not satisfy (9), but the noncyclic one (namely, S_3) is a subgroup of A_5: it can be generated, for example, by (123) and (12)(45). All groups of order 8, except Z_2^3, have elements of order 4 and do not satisfy (9).

2. The group Z_2^3 requires more effort. First we note that if

$$a = (a_1a_2)(a_3a_4) \text{ and } b = (b_1b_2)(b_3b_4)$$

are two elements of order 2 in A_5, then there are three possibilities:

(i) $\operatorname{Supp}(a) = \operatorname{Supp}(b)$, say $a = (12)(34)$ and $b = (13)(24)$. Then $ab = ba = (14)(23)$, so a and b are contained in the same Sylow 2-subgroup of A_5.

(ii) $|\operatorname{Supp}(a) \cap \operatorname{Supp}(b)| = 3$ and a and b have a common cycle, say $a = (12)(34)$ and $b = (12)(35)$. Then $ab = (345)$, so $|ab| = 3$.

(iii) $|\operatorname{Supp}(a) \cap \operatorname{Supp}(b)| = 3$ and a and b have no common cycles, say $a = (12)(34)$ and $b = (15)(24)$. Then $ab = (14325)$, so $|ab| = 5$.

Now consider the following formula in three variables x_1, x_2, x_3:

$$\Big[\bigvee_i x_i^{15}=1\Big]\vee\Big[\bigvee_{i<j}(x_ix_j)^{15}=1\Big]\vee\Big[\bigvee_{(i,j,k)\neq 0} x_1^ix_2^jx_3^k=1\Big] \tag{10}$$

where the last disjunction is taken over all nonzero $\{0,1\}$-vectors (i,j,k). Take any x_1, x_2, $x_3 \in A_5$. If the first two clauses of (10) are false then all x_i are elements of order 2 belonging to the same Sylow 2-subgroup H of A_5. Since H is a 2-dimensional vector space over $GF(2)$, the last clause must hold. Thus (10) is valid in A_5. On the other hand, it is obviously false in Z_2^3.

Let G be any group whose order is divisible by 8 and which satisfies (9). Then G has a subgroup H of order 8; in view of (9), $\exp H = 2$ and so $H \simeq Z_2^3$. This eliminates all groups of orders 16, 24, 32, 40 and 48.

3. There are two groups of order 9: Z_9 and $Z_3 \times Z_3$. The first does not satisfy (9). To eliminate the second, take the formula

$$(x_1^{10}=1)\vee(x_2^{10}=1)\vee(x_1=x_2)\vee\big((x_1x_2)^{10}=1\big)\vee\big((x_1^2x_2)^2=1\big). \tag{11}$$

If x_1, $x_2 \in A_5$ and $x_i^{10} \neq 1$, then x_i are cycles of length 3. Without loss of generality we assume that $x_1 = (123)$. If x_2 moves the same symbols 1, 2, 3, then either $x_1 = x_2$ or $x_1x_2 = 1$. If x_2 moves only one of the symbols 1, 2, 3, then x_1x_2 is a cycle of length 5 and $(x_1x_2)^5 = 1$. If x_2 moves two of the symbols 1, 2, 3, then there are two essentially different cases: $x_2 = (124)$ and $x_2 = (214)$. In the first case

$$x_1x_2 = (123)(124) = (14)(23) \text{ and so } (x_1x_2)^2 = 1,$$

in the second case

$$x_1^2x_2 = (132)(214) = (13)(24) \text{ and so } (x_1^2x_2)^2 = 1.$$

It follows that (11) is valid in A_5. Clearly it is not valid in $Z_3 \times Z_3$, and so the groups of order 9 are eliminated. This also eliminates all groups of orders 18, 27, 36, 45 and 54.

4. There are two groups of order 10: Z_{10} and D_{10}. The first does not satisfy (9) but the second is a subgroup of A_5. There are five groups of order 12: Z_{12}, $Z_2 \times Z_2 \times Z_3$, A_4, D_{12} and also the group $H = \langle a, b \mid a^3 = b^4 = 1, a^b = a^{-1}\rangle$. All of them except A_4 have elements of order 6 and therefore do not satisfy (9), while A_4 is a subgroup of A_5. The only group of order 15 is cyclic. Every group G of order 20 has only one Sylow 5-subgroup, and so $G = Z_5 \rtimes T$ where $|T| = 4$. Since G may not have elements of order 4, we have $T = Z_2 \times Z_2$. But then $C_T(Z_5) \neq 1$, and so G contains an element of order 10, which again contradicts (9). Thus the groups of order 10, 12, 15 and 20 are eliminated.

5. Every solvable group of order $15m$ with $(15, m) = 1$ contains a subgroup of order 15. This eliminates all solvable groups of order 30 and 60, but the only nonsolvable group of order ≤ 60 is A_5 itself.

6. To eliminate the groups of order 25, consider the formula

$$\begin{aligned}(x_1^6=1)\vee(x_2^6=1)\vee\big((x_1x_2)^6=1\big)\\ \vee\big((x_1x_2^2)^6=1\big)\vee\big((x_1x_2^3)^6=1\big)\vee\big((x_1x_2^4)^6=1\big).\end{aligned} \tag{12}$$

It is valid in A_5. Indeed, take x_1, $x_2 \in A_5$ with $x_i^6 \neq 1$. Then the x_i are 5-cycles. It is easy to verify the following general property of the symmetric groups: *if α and*

β are two n-cycles in S_{n+1}, then for some $d \leq n-1$ the permutation $\alpha\beta^d$ is not an n-cycle. In our case, there exists d ($1 \leq d \leq 4$) such that $x_1x_2^d$ is not a 5-cycle, that is $(x_1x_2^d)^6 = 1$. On the other hand, (12) fails in any group of order 25: take an element $x_2 \in G$ of order 5 and any element x_1 not in $\langle x_2 \rangle$. This eliminates the groups of order 25 and 50 and completes our argument.

THEOREM 3. *Formulas ω_{60} and (9)–(12) form a basis of d-identities of A_5.*

Let us now find a basis of identities of $\operatorname{Reg} A_5$, provided $\operatorname{char} K \neq 2, 3, 5$. First, we apply the "*-operation" to the formulas ω_{60} and (9)–(12) and obtain five elements of KF. Next, we add two more elements of KF:

$$x^{30} - 1 \tag{13}$$

and the standard polynomial of degree 61

$$s_{61} = \sum_{\sigma \in S_{61}} (-1)^\sigma x_{\sigma(1)} \dots x_{\sigma(61)}. \tag{14}$$

This set of seven elements of KF is a basis of identities of $\operatorname{Reg} A_5$. Indeed, both (13) and (14) are valid in $\operatorname{Reg} A_5$, because $\exp A_5 = 30$ and because $s_{n+1} \equiv 0$ in any algebra of dimension n. In view of Lemma 4, it remains to show that the variety defined by (13) and (14) is locally finite and ordinary. Let $\rho: G \to \mathrm{GL}(V)$ be a faithful representation satisfying both (13) and (14). We may assume that $G \subset \operatorname{End} V$, and then $s_{61}(g_1, \dots, g_{61}) = 0$ for any $g_i \in G$. Let A be a subalgebra of $\operatorname{End} V$ generated by G. Since s_{61} is a multilinear polynomial, it follows that $s_{61} \equiv 0$ in A. Hence G is a periodic subgroup of a PI-ring A. It is known that such group is locally finite (see, for example, [**Pr**]). Its exponent must divide 30, and so ρ is ordinary. This concludes the consideration of Example 2.

Note that the above argument works for any group G. Indeed, let $\{\delta_i\}$ be a weak basis of d-identities of G. If $\exp G = e$ and $|G| = n$, then the set $\{\delta_i^*\}$ plus $x^e - 1$ and s_{n+1} is a basis of identities of $\operatorname{Reg} G$.

It is more difficult to find a basis of d-identities of A_6. We state the final result; the proof will be published elsewhere (see [**V5**]).

THEOREM 4. *The following formulas form a basis of d-identities of A_6:*

$$\omega_{360} = \bigvee_{0 \leq i < j \leq 360} (x_i = x_j),$$

$$(x^3 = 1) \vee (x^4 = 1) \vee (x^5 = 1),$$

$$\Big[\bigvee_{1 \leq i \leq 9} x_i^{20} = 1\Big] \vee \Big[\bigvee_{i<j} x_i = x_j\Big] \vee \Big[\bigvee_{i<j} (x_ix_j)^{20} = 1\Big] \vee \Big[\bigvee_{i<j} (x_ix_j^2)^{20} = 1\Big],$$

$$(x_1^{30} = 1) \vee (x_2^{30} = 1) \vee ((x_1x_2)^{15} = 1) \vee ((x_1x_2^2)^{15} = 1) \vee ((x_1x_2^3)^{15} = 1),$$

$$\theta(x_1, x_2, x_3) \vee \Big[\bigvee_i x_i^{15} = 1\Big] \vee \Big[\bigvee_{i \neq j} (x_ix_j)^{15} = 1\Big] \vee \Big[\bigvee_{i \neq j} (x_ix_j^2)^{15} = 1\Big] \vee \Big[\bigvee_{i \neq j} (x_ix_j^3)^{15} = 1\Big],$$

$$(x_1^{12} = 1) \vee (x_2^{30} = 1) \vee ((x_1x_1^{x_2})^{12} = 1) \vee ((x_1x_1^{3x_2})^{12} = 1) \vee ((x_1^2x_1^{x_2})^{12} = 1),$$

$$(x_1^{12} = 1) \vee (x_2^{12} = 1) \vee \big((x_1x_2)^{12} = 1\big) \\ \vee \big((x_1x_2^2)^{12} = 1\big) \vee \big((x_1x_2^3)^{12} = 1\big) \vee \big((x_1x_2^4)^{12} = 1\big).$$

Here $\theta(x_1, x_2, x_3)$ *denotes the formula* (5).

The considerations of this section demonstrate some general principles for computing a basis of d-identities of a finite group G. If $|G| = n$, then we first take ω_n, and then for every group H of order $\leq n$, which is not a section of G, find a d-identity of G that is not valid in H. It is clear that the complexity of this procedure increases dramatically with the order of the group in question. In general, it is noticeably easier to find a *weak* basis of d-identities, because in this case we do not have to think about groups not admitting faithful irreducible representations. In particular, we do not have to think about groups of the form Z_p^n, which usually create most of the trouble. It is not excluded that some computer-based system (like GAP or CAYLEY) could be involved in this process, and we hope to investigate such possibilities in the future.

References

[B] K. A. Baker, *Equational axioms for classes of lattices*, Bull. Amer. Math. Soc. (N.S.) **77** (1971), 97–102.

[BO] Yu. A. Bakhturin and A. Yu. Ol'shanskii, *Identical relations in finite Lie rings*, Mat. Sb. **96** (1975), 543–559.

[Br] R. Bryce,, *Metabelian groups and varieties*, Proc. Roy. Soc. London Ser. A **266** (1970), 281–355.

[G] G. Grätzer, *Universal algebra*, 2nd ed., Springer, 1979.

[I] A. V. Il'tyakov, *On finite bases of identities of Lie algebra representations*, Nova J. Algebra Geom. **1** (1992), 207–259.

[K1] A. N. Krasil'nikov, *On the finite basis property of the identities of nilpotent-by-abelian groups*, Izv. Akad. Nauk. SSSR Ser. Math. **54** (1990), 1181–1195.

[K2] ———, *On the identities of representations of groups by triangular matrices over a commutative ring*, Contemp. Math. **131** (1992), 217–225.

[Kr] R. L. Kruse, *Identities satisfied by a finite ring*, J. Algebra **26** (1973), 298–318.

[L] I. V. L'vov, *On varieties of associative rings. I.*, Algebra i Logika **12** (1973), 269–297.

[Mc] R. McKenzie, *Equational bases for lattice theories*, Math. Scand. **27** (1970), 24–38.

[N] H. Neumann, *Varieties of groups*, Springer, 1967.

[O] A. Yu. Ol'shanskii, *The soluble just-non-Cross varieties of groups*, Mat. Sb. (1971), 115–131.

[OP] S. Oates and M. B. Powell, *Identities of finite groups*, J. Algebra **1** (1964), 11–39.

[P] B. I. Plotkin, *Locally finite and locally finite-dimensional varieties of pairs—group representations*, Collection of Papers in Algebra, Riga, 1978, pp. 185–245.

[PK] B. I. Plotkin and A. H. Kushkuley, *Identities of regular representations of groups*, preprint.

[PV] B. I.Plotkin and S. M. Vovsi, *Varieties of group representations: general theory, connections and applications*, Zinatne, Riga, 1983.

[Pr] C. Procesi, *The Burnside problem*, J. Algebra **4** (1966), 421–425.

[R1] Yu. P. Razmyslov, *On a problem of Kaplansky*, Izv. Akad. Nauk. SSSR Ser. Mat. **37** (1973), 483–501.

[R2] ———, *Identities of algebras and their representations*, Nauka, Moscow, 1989.

[V1] S. M. Vovsi, *On critical representations of groups*, Latv. Mat. Yezhegodnik **20** (1976), 141–159.

[V2] ———, *Triangular products and identities of certain associative algebras*, Algebra i Logika **26** (1987), 27–35.

[V3] ———, *Topics in varieties of group representations*, London Math. Soc. Lecture Note Ser., vol. 163, Cambridge Univ. Press, Cambridge, 1991.

[V4] ———, *On the identities of representations of finite groups over commutative rings*, Internat. J. Algebra Comput. **2** (1992), 103–116.

[V5] ———, *Disjunctive identities of finite groups and identities of regular representations*, Algebra Colloq. **4** (1997), 257–280.

[VN] S. M. Vovsi and Hung Shon Nguyen, *Identities of stable-by-finite representations of groups*, Mat. Sb. **132** (1987), 578–591.

[Z] E. I. Zelmanov, *On Engel Lie algebras*, Sibirsk. Mat. Zh. **29** (1988), 112–117.

Department of Mathematics, Rutgers University, New Brunswick, NJ 08903, USA

E-mail address: vovsi@math.rutgers.edu